YAMAHA

YAMAHA

ALL FACTORY AND PRODUCTION ROAD-RACING TWO-STROKES FROM 1955 TO 1993

Colin MacKellar

The Crowood Press

First published in 1995 by
The Crowood Press Ltd
Ramsbury, Marlborough
Wiltshire SN8 2HR

www.crowood.com

This impression 2008

© Colin MacKellar 1995

All rights reserved. No part of this publication may be reproduced or transmitted in any form or by any means, electronic or mechanical, including photocopy, recording, or any information storage or retrieval system, without permission in writing from the publishers.

British Library Cataloguing-in-Publication Data

A catalogue record for this book is available from the British Library.

ISBN 978 1 85223 920 6

Picture Credits

The photographs on pages 15 and 38 are by Sonny Angel; those on pages 17 (top), 19 (bottom), 21, 41 (top) and 43 are by Autoby; those on pages 27, 28 (top), 47 and 52 are by Daytona International Speedway; those on pages 24 (top), 29 (top), 31, 34, 35, 36, 54 (bottom), 55 (bottom), 56, 57, 58 (bottom), 61, 63 (bottom), 64, 67 (top), 68 (all), 78, 80, 82 (all), 86, 87, 88, 89, 93, 96 (top), 99, 105, 106, 107, 108, 115, 116, 117, 124, 126, 127, 136, 149, 152, 153, 154, 156, 158, 160 and 162 (all) are by Jan Heese; those on pages 128, 130, 131 (all), 133, 134, 137, 138 (all), 140, 142, 146 (bottom), 147, 163, 164, 167 (all), 168, 169, 170, 171, 172, 173, 174, 175, 176 (all), 179 and 180 are by Henk Keulemans; that on page 29 (bottom) is by Shigeo Kibiki; that on page 113 is by Dan Mahoney; those on pages 22, 32, 41 (bottom), 45 and 54 (top) are by Nic Nicholls; those on pages 120, 143 (all) and 145 are by John Owens; that on page 112 is by Bert Shepard; those on pages 30, 74, and 88 are by Mick Woollett; those on pages 9, 11, 13, 19 (top), 28 (bottom), 37 and 40 (all) are by Yamaha.

Colour section photographs by Jan Heese, except those on pages 5 (top), 7 (bottom left) and 8, which are by Henk Keulemans, and those on pages 5 (middle), and 6, which are by Yamaha.

Printed and bound in Great Britain by The Cromwell Press, Trowbridge

Contents

	Preface	6
	Acknowledgements	7
1	How the West Was Won First Factory Racers 1955–62	9
2	Works Glory Factory Racers 1962–8	22
3	Growing Pains TD1 Production Racers 1962–8	37
4	Heir to the Throne TD/TR Production Racers 1969–73	49
5	Private Passion TZ Racers 1973–80	62
6	The Crown Jewels Piston-Ported 500s 1973–81	78
7	The Beast TZ750 1973–9	101
8	The Tenacious Twins TZ250 1981–93	121
9	The Doldrums Factory 500s 1981–3	149
10	Biggest Bang per Buck Factory Reed-Valve V4s 1984–93	163
	Appendix	184
	Index	189

Preface

I believe the first racing motorcycle I ever saw in action was a TZ250 Yamaha. It was a total sensory experience starting with a buzz in the middle distance as I removed my helmet in the grass field that doubled as a bike park at Brands Hatch, England back in 1976. It was the Friday of a three-day Easter weekend which I had decided to kick off by watching Sheene and company thrash the Yanks at the first round of the year's Trans-Atlantic Trophy. As I moved towards the circuit entrance, I was hit by the acrid smell of racing two-stroke exhaust fumes, which got stronger as I hurried along the road behind the stands under the Dunlop bridge. Pushing past the fanatics who'd arrived even earlier than me, I caught my first glimpse of the circuit, at the end of the start/finish straight at the top of the renowned off-camber downhill right-hander called Paddock Bend. The sight of the droves of TZ250s out for early morning practice was totally electrifying, firing a passion for racing that remains with me today. Even the thrashing handed out to the British by the Americans, in particular by S. Baker, K. Roberts and G. Nixon, couldn't spoil the magic of my first visit to a racetrack. I returned to Brands many times that and subsequent years, the magic never seeming to wane.

The TZs, and more generally, Yamaha's contribution to the world of motorcycle racing is staggering. The recent celebration of thirty years of 250cc TD and TZ production racers is unparalleled in the sport, representing a statement of commitment to racing that goes far beyond the commercial process of building and selling motorcycles. Yamaha mass produced the first over-the-counter racing superbike, which was to become the icon of US racing during the 1970s. The TZ750 was the tool with which the first wave of US riders honed their skills for the European racetracks that formed the heart of Grand Prix racing. Alone among motorcycle manufacturers, Yamaha answered the pleas from the racing world to save the 500cc class of GP racing from a slow death by haemorrhage. The Harris and ROC Yamahas stemmed the bleeding; the jury is still out on the health of the patient. Over 350 GP victories have been Yamaha's reward for their support of the sport as well as a place in the hearts of many of those who observe and participate in the world of racing.

Acknowledgements

I approached the task of documenting this epic with mixed emotions, excited at the opportunity to tap into the racing community, but concerned that changes during the last ten years no longer made it possible. My concerns were groundless. Despite the massive commercialization that the GP world has undergone, it is still possible to gain access to the men around which the business has been constructed. Paul Butler and Mike Trimby of IRTA and Dennis Noyce of Dorna let me trawl the paddock at the 1992 and 1993 Dutch TT, interviewing Kenny Roberts, Kel Carruthers, Giacomo Agostini, Erv Kanemoto, Vince French, Warren Willing, Bud Aksland and Wayne Rainey in the process. I'm very grateful to all of them for making time available during the nerve-racking days before a GP. Thanks also to journalist Michael Scott for pulling out his little black book and providing telephone numbers of many involved in Yamaha's racing history.

Others provided valuable insights into Yamaha teams and machines of the past. Ferry Brouwer relived his days and nights at the track with Phil Read and Bill Ivy in 1968, Phil Read in 1970 and 1971, Chas Mortimer in 1972 and Jarno Saarinen and Tepi Lansivouri in 1973. Vince French talked of his time with Saarinen, Lansivouri and Cecotto, whilst Mac MacKay recalled life in Team Yamaha as Agostini fought for his last and Yamaha's first 500cc title. Uniquely, Trevor Tilbury could look back at more than 15 years working with Yamaha, most of them with Kenny Roberts, whose GP team he co-ordinates today.

A book like this can only be as good as its information sources and I was fortunate in obtaining the help of many knowledgeable and helpful people. Ludi Beumer was of enormous assistance in supplying parts books and service manuals for so many of the production racers as well as acting as go-between with the Yamaha PR department in Japan. Guus ten Thije provided articles and service sheets and tore down his 1991 TZ250B for my scrutiny. Kevin Cameron responded to my request for porting details with a comprehensive set of measurements of the TZ250 through the 1980s and 1990s. Rob Bron helped me locate some of the missing data on TZ250 specifications (in the Appendix) as did Dennis Trollope, Bob Eaves and Dave Hedison. Jim Reed of the now-defunct Yamaha T-register deserves a word of thanks for the excellent work he did in assembling data on Yamaha twins: the newsletters are still a valuable source of information.

The challenge of providing excellent photographs to illustrate the book was eased by the enthusiasm shown by Jan Heese and Henk Keulemans for the project. It was a pleasure to work with them to make a selection, the quality of their material sometimes making it agonizingly difficult to make a choice. I am also especially grateful to Mr. Toyama at Yamaha for providing me with photographs from Yamaha's early racing history in Japan, and Shigeo Kibiki for locating and sending material from the Autoby archive. Other photographs were provided by John Owens, Daytona Speedway, Bert Shepard, Dan Mahony, Nic Nicholls and Mick Woollett.

Finally I would like to thank those who

Acknowledgements

were directly instrumental in bringing the book to publication. Tony Pritchard initially approached me with the proposal, but was unable to complete the publication process. Ken Hathaway at Crowood took the book on and nurtured it through to completion. Tim Parker at Motorbooks International was prepared to make time available to discuss the project and offer valuable advice when requested. Marian, Cara and Niall did their best to create the space I needed to get the job done, sometimes at great inconvenience to themselves. Jongens bedankt! En Niall, ik ga echt niet zo vaak meer boven typen.

1 How the West Was Won
First Factory Racers 1955–62

On 9 July 1955, the third Mount Fuji Climbing Races were held under the auspices of the Tokyo Motorcycle Race Association. This organization was formed round a group of motorcycle dealers from Tokyo who had decided two years before to stimulate the Japanese motorcycle industry by organizing race meetings open solely to 100 per cent Japanese machines. Every single component had to have been manufactured locally down to the last nut and bolt. Pre-war racing had been the domain of the foreign motorcycles and the post-war Japanese economy had not recovered sufficiently to enable racing of any sort to be considered. The race at Fuji was won by a 90cc Auto-bit. Later that year the Nagoya TT was organized, attracting entries from nineteen

A Red Dragonfly (YA1) disguised as an Iron Horse leads the parade of riders for the first Asama Plains race in November 1955.

manufacturers, and it developed into a two way battle between the ohc Showas and the three-valve ohv Hondas. Showa won the race by 18 seconds after four hours of racing on the 146 mile (235km) course over largely unpaved roads. The 1954 Mount Fuji race was to be the Monarch's hour of glory in the 250 class, as both Honda and Showa were soundly beaten by the Monarch's average speed of 34mph (55km/h). The Suzuki Jidosha Kogyo company won the 90cc race and went on to greater things, although repetition of this success was to elude them for many years to come.

Then came the third race on the slopes of Japan's Holy Mountain, over tracks consisting of compacted volcanic ash, overgrown in places and deeply rutted along its entire length. Moves to set up a more formal governing body to co-ordinate the development of racing in Japan had resulted in less interest in the 1955 Fuji race and only twelve marques were represented at the startline for the 90cc and 250cc races that were to be run concurrently. One of these marques had not been racing before; their name was Yamaha Motor.

FIRST RACERS: YA AND YD

The company was barely a year old and their motorcycle, the YA1 'Red Dragonfly', less than six months old. The company had been formed by parent Nippon Gakki when they had decided to diversify their business interests that had until that time been based primarily on the manufacture of musical instruments. It is unclear whether it was business shrewdness or just good luck that caused Yamaha to enter the motorcycle business when they did. A massive shake-out of the industry had been going on for some years and already more than 25 per cent of the forty motorcycle manufacturers that had been based at Hammamatsu in 1952 had disappeared. By coming into the business at this late stage, Yamaha were able to analyse the cause of the failures of these companies and use the insights thus gained to produce a successful competitor to the market dominated by Tohatsu, Cabton and Pointer. No doubt they must also have been encouraged by the aggressive 'Buy Japanese' campaign announced by the government in 1954, that was backed up by an 80 per cent import and commodity tax imposed on foreign motorcycles. Protectionism was not such a dirty word in those days. The one exception to this crippling tax was the waiver extended to Japanese companies who wished to purchase foreign motorcycles with a view to 'studying' them. They could even borrow money from the government to finance the 'study'. Yamaha requested and were granted permission to import a DKW RT125 for this purpose. The design phase of the YA1 was as good as completed once the DKW had been uncrated at Hammamatsu.

No one can blame Yamaha for taking this plagiaristic approach to break into the motorcycle manufacturing business. With no experience or expertise in automotive engineering and insufficient resources to finance a big research and development programme to design their own machine, the active encouragement of the government made it an obvious choice. The genius on their part was to choose a German machine to copy. The companies that used German machines as models, notably Suzuki and Honda, went on to success alongside Yamaha. Many of the others used British machines and preceded the British industry into decline and fall.

The YA1 was a 207lb (94kg) featherweight that developed 5.6bhp from its 125cc engine measuring 52 × 58mm turning at 5,500rpm. A 20mm Amal carburettor clone was fitted, considered large for the time, and the four-speed transmission gave it a top speed close to 50mph (80km/h). This was heady stuff for a Japanese 125 machine, as Suzuki's 100cc

Two of the factory YA1s that did so well at the first Asama race.

Porter Free was barely capable of 35mph (55km/h). The Yamaha development staff managed to put their personal stamp on the machine by designing the four-speed gearbox, while the DKW only used three gears. They also made contact with the Tokyo Academy of Arts to bring in Professor Koike to act as industrial design consultant, marking the start of a relationship that would flourish and continue into the 1980s.

By July 1955, with almost 1,000 units produced, Yamaha were ready to demonstrate their own perceived superiority on the racetrack. On 9 July 1955, a team of three riders lined up for the start of Yamaha's debut motorcycle race. It is doubtful if any motorcycle manufacturer has ever had such a convincing win in its first competition event. Riding essentially standard Dragonflys with knobbly tyres, the three rushed into the lead and were never to be headed. One of the machines failed with just 3 miles of the 17 mile circuit to go, but OishiSan headed a Yamaha 1–2, finishing in just over 25 minutes, 4 minutes less than the previous year's fastest time and four minutes before the third placed Showa. Yamaha management were rightly jubilant, sales enquiries poured in, waiting lists developed for the YA1 and all was right in the world for the company.

To show that it was no fluke and to reinforce their position as lightweight class champion, Yamaha signed up for the forthcoming Asama race scheduled to be held on 5 November. They entered four riders under the management of WataseSan, whose dedication to pre-race preparation and

How the West Was Won

The Adler which Yamaha 'studied' before building their first twin, the YD1.

competitive analysis have earned him a position as one of the legends in Yamaha's racing history. It was he who had the Yamaha team out on the circuit practising at 5.30 a.m. before the other teams were even out of bed. The riders had good knowledge of the circuit by the time the race got underway at 12.00 p.m. in near-freezing conditions. The race was to be a repeat of the total domination that the YA1s had achieved at Fuji. All four machines finished the three-lap race, filling the first four places and beating both Suzuki and Honda in the process. The Honda effort in particular was a complete fiasco despite a big effort from the factory. Two machines retired and the other two took ninth and tenth places. Yamaha were the undisputed kings of the lightweight class in Japan.

Despite the great success of the races at Asama in 1955, two years were to pass before another race meeting was held. The main reason for this was the construction of an unpaved closed circuit on the slopes of Mount Asama, to be devoted solely to motorcycle racing. The circuit was intentionally left unpaved since most of the roads in Japan at that time were not asphalted, and the circuit was intended to function at least partially as testing ground for the street legal machines. Total length of the course was now just over 5½ miles (9km).

In October 1957 the second Asama race was held and Yamaha were able to field not only a 125 team but also a 250, for in April 1957 they had unveiled their first 250, the YD1. The bike that had been 'studied' before the YD1 was produced was the Adler MB250, but in fact the design team was scarcely influenced by the German twin. The 54 × 54mm twin displaced 247cc and produced 14.5bhp at 6,000rpm. Weighing in at 309lb (140kg) the twin would have been considered light for its time, despite the use of a

competitive in international competition and a milestone in the racing history of the company had been laid.

After Catalina, the team returned to Japan but the bikes did not. Sonny Angel was given the YD-As and entered a young rider by the name of Calvin Rayborn in a number of half-mile races. The wins he achieved on the bike were an indication of great things to come.

Back in Japan, racing continued with the first clubman's race at Asama open to any domestic or international machines based on street bikes. Yamaha's YD1 racers were illegal, as they bore little relationship to the utilitarian YD1. Consequently Yamaha did not contest the 250 class. They wiped up as usual in the 125 class taking the first five places on a rain-swept course on 29 August 1958. They also did surprisingly well in the 350 race, in which a 260cc YE1 finished in second place, a long way behind a BSA Gold Star ridden by future Honda rider Kunimitsu Takahashi. The YE1 was simply a bored out YD1 and had been announced the previous month. That it was able to beat the Honda C75Z, 305cc ohc twins says more about the Hondas than it does about the Yamaha.

It is unclear what exactly happened at Honda during the winter of 1959–60, but something or someone galvanized them into full commitment to racing on a grand scale. Until that winter, Yamaha were perceived to be the leaders of the Japanese manufacturers participating in racing, but 1959 was to mark the start of success for Honda inside and outside Japan that was to leave Yamaha struggling to catch up throughout the 1960s. They seemed to slip from a position of leader

Calvin Rayborn (4) on his way to victory in the El Cajun half-mile TT in May 1959.

to follower, from a pro-active style to a reactive style, matching Honda's moves, but only occasionally causing Honda to match their own. Honda's frustration at their failure to successfully race modified street machines, led them to reject this approach in favour of the design of hi-tech racers, whose development might take years to filter down to the street machines. Yamaha continued to support a strong link between racers and street bikes right through to the 1980s, only building complex factory racers when no other alternative was possible. This philosophy served them well in the long term, but it caused them to live in Honda's shadow for the next five years of GP racing.

On Saturday, 22 August 1959, the last big Asama race meeting got underway. It was to prove a huge success for Honda. Clubman and full international races were held, both being open to Japanese and foreign machines. C90 Hondas thrashed the suddenly old YAs in the 125 race. A YDS1R managed to take third place in the 250 event and Tanahara Noguchi brought a YES1R home in first place, 1.5 minutes ahead of the first Honda. The next day, the unrestricted race machines were to compete and it saw the historic debut of the Honda RC160 four-cylinder 250. This was essentially a doubled-up version of the RC142 that the factory had used to win the Isle of Man team award two months before. Yamaha had withdrawn from the 125 and 250 races, but Noguchi decided to enter his YDS1R anyway. The 250 race got underway on the Sunday afternoon, in poor racing conditions. Dense fog was swirling around the course, reducing visibility to tens of metres in places. The three RC160s rushed into lead, setting a lap record of 106km/h in the process, but Noguchi was not far behind despite his 29lb (13kg) and 15bhp disadvantage. The muddy circuit was making the powerful Hondas quite a handful and their riders were having difficulty getting the power to the ground.

On the seventh lap Noguchi managed to pass the third place Honda ridden by Giichi Suzuki. For four laps he managed to fend off the Honda but was finally passed on the main straight along the start/finish area. One lap later Noguchi was out with a broken transmission main shaft and Honda comfortably took the first three places. Noguchi completed a good weekend by taking the 350 race on his YES1R by more than 10 minutes!

This marked the end of an era of Japanese volcano racing. Honda went off to seek their fame and fortune in Europe and GP racing became the focus of attention in Japan. Yamaha were prevented from joining them there by an acute lack of money caused by a disastrous assault on the scooter market and the acquisition of the Showa motorcycle division, together with Hosk who had themselves been purchased by Showa in 1959. The 125 Showa machines had been fast at the Asama race and looked like having the potential for GP winning performance, if only their reliability problem could be fixed. Yamaha decided to work to a limited trip to Europe during 1961 to test the waters using machinery based on the Showa 125 machine with disc valve induction.

Towards the end of the 1950s, disc valve induction was considered to be the 'great white hope' for two-stroke engine designers in their battle for four-stroke power and reliability. As is so often the case, it was experimentation by a German company that had led to the disc valve becoming the panacea to cure all two-stroke engine evils. Walter Kaarden, working for MZ, had started building disc valve engines in 1953 and by 1955 they were winning international races. Kaarden's enthusiastic embrace of the disc valve arose from the advantage it gave with inlet timing. With inlet timing performed by the piston skirt, the port would always open and close symmetrically around TDC. Opening early to flow plenty of charge would mean closing late and losing it back out of

The RA41 looked quite businesslike in the confines of the factory but was totally outclassed by contemporary 125cc GP machines.

the carburettor. At the time, there was little understanding of the resonances active in both the inlet and outlet tract, and which are harnessed to such great effect on today's two-strokes. Carburettor technology was 'stone-age'. The disc valve would allow the inlet port to the crankcase to open as early as required and could be closed as soon as the danger zone of blowback approached. MZ began getting GP decent power from their two-strokes by the late 1950s, especially after adding extra transfers at the back of the cylinder where the inlet port would be on a piston-porter. Ernst Degner's MZ win at Monza in September 1959, the last 125 GP of the decade, marked the death of the four-stroke in GP racing as sure as the start of the Ice Age marked the end of the dinosaurs.

Showa had been moving along in the same general direction as MZ, but were still a couple of years behind in development. Hata was given the task within Yamaha of transforming the Showa into GP winning material for the 1961 planned visit to Europe. The result was the RA41 air-cooled single cylinder six-speed machine. Measuring 56 × 50mm, it was a fairly

The black-tanked RD48 carried Yamaha's colours to Europe but was not a great success.

Oishi (4) makes a smokey start to the first GP contended by Yamaha, the French at Clermont-Ferrand.

conventional engine in design, with massive cooling fins on the aluminium cylinder that was chrome plated, rather than being anodized or running with an iron-liner. The thermal bonding between liners and cylinder muffs had always been a problem in engine design and the anodization used on the YDS1R had removed one potential source of seizure. Unfortunately, the anodized surface of the cylinder wall was not good at retaining the thin oil film required for long life and happiness in a two-stroke engine. Experimentation by the factory had led them to choose a porous chrome coating to reduce the risk of seizure even further.

There were two unusual aspects to the engine. Firstly it was fitted with two 24mm carburettors, one on either side of the engine. This was done in an attempt to get more charge into the crankcase as well as to provide even cooling to the crankshaft bearings. In fact, bearing lubrication was a major worry to them as they supplemented the petroil lubrication from the charge with forced lubrication of crankshaft and big end by an oil pump mounted on top of the gearbox. The major deficiency of the engine was its lack of a third transfer port on the rear wall of the cylinder. MZ still had the march on them as far as two-stroke technology engineering was concerned. Housed in a conventional cradle frame, the RA41 produced just under 20bhp at 10,000rpm and weighed in at a respectably light 190lb (85kg).

The RD48 was a doubled up RA41, using the same chassis, fuel tank and seat. It had 32mm carburettors fitted to each side of the crankcase, feeding each 56×50mm cylinder via the disc valve splined to the outboard end of each crankshaft. An Oldham coupler was used to join the inboard end of each crankshaft. A primary gear on the line of the vertically split crankcase, drove the primary transmission gear mounted on a jackshaft above and behind the crankshafts. The jackshaft drove the small oil pump providing

Noguchi waits nervously for Isle of Man practice to start for the 125s and his RA41.

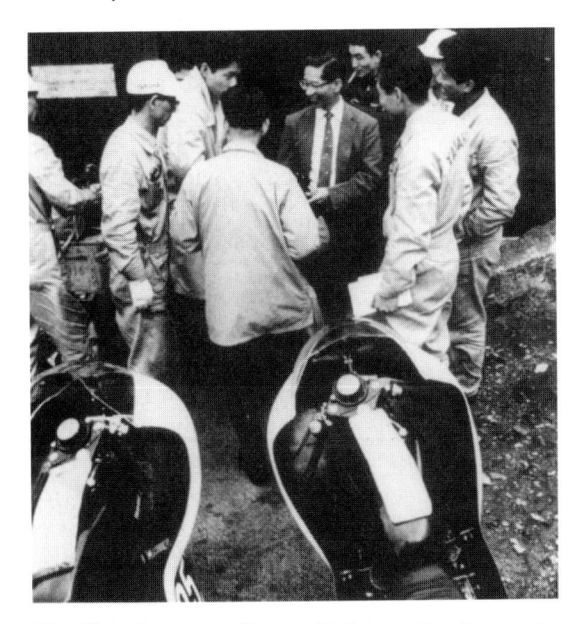

The Yamaha team discuss Ito's practice times at the Island, with two RA41s in the foreground. The mechanic on the right with his hands in his pockets is Toru Hasegawa, Yamaha's current president.

pressured lubrication of the crankshaft, via a worm gear. It also drove the dry clutch via a small gear on its right outboard end and the racing magneto via the same primary gear. Again Yamaha had managed to keep the weight down and the 35bhp the engine produced gave the 220lb (100kg) machine a reasonable turn of speed.

The 1961 GP season began and Yamaha planned to race at the French GP at Clermont-Ferrand, the Isle of Man, Assen and Spa. It threatened to go horribly wrong before they even saw a European race track. The bikes arrived late in France and customs clearance took longer than the team had expected. While the riders spent their time admiring the Palace of Versailles and the Eiffel tower, team manager Naito and bursar Miyake spent three days getting everything officially imported into Europe. With all the paperwork completed, the team rushed off to Clermont-Ferrand and arrived just as practice for the 250 was about to start. Not being able to find their racing numbers, Ito quickly painted a rough 36 on his fairing whilst Noguchi was ushered numberless onto the circuit. In fact the whole French GP was not a great success. Yamaha opened their GP racing career on Sunday, 21 May 1961, with an eighth place by Noguchi in the 125 class and another eighth place by Ito in the 250. Not exactly a glorious debut.

Things could only get better and they did. At the Isle of Man, Gary Hocking, the Rhodesian riding for MV, graciously helped Ito and Noguchi learn the circuit and Ito managed to earn himself a single World Championship point with a sixth place on the RD48. For the first time ever, all the Championship points for this race were earned by Japanese machines, five Hondas and the single Yamaha. The 125 was less of a success, with Ito taking eleventh place from team-mate Oishi. On to Assen where Ito repeated his sixth place on the RD48 and

19

How the West Was Won

Ito (36) and Sunako (4) get underway for Yamaha's last 250 GP of 1961 at Spa. They managed fifth and sixth places.

was supported by Noguchi who was pipped on the line for seventh place. Sunako and Ito managed eighth and ninth places respectively in the 125, whilst local Dutch rider Cees van Dongen could only manage twelfth on his loaned RA41. Finally a week later at Spa, Sunako spluttered home in thirteenth place on the 125, drawing to a close a bitterly disappointing venture in 125cc GP racing. Luckily, Ito's fifth and Sunako's sixth place in the 250 race compensated for the disastrous 125 result. The RD48 had proved to be a fairly successful machine, having garnered a total of five Championship points to its name. But closer examination of the results of the Belgian GP reveal how far off the pace they really were. Ito finished over two minutes later than the fourth placed Honda of Shimazaki and almost five minutes behind race winner Jim Redman. This must surely have given Yamaha food for thought on their long journey back to Japan. They were to race one more time that year, at the Argentinian GP in October. Ito finished last in the 250 race which totalled four finishers, so another three points were added to Yamaha's total for the year.

The 1961 expedition to the GP world had been an invaluable experience for the factory. Ironically, after their success during the 1950s, the 125 had been a totally uncompetitive machine. Although engine and chassis were light it had been seriously underpowered, underbraked and overfaired. The enormous fairings had the aerodynamic qualities of a Sherman tank and were heavy to boot. They were less of a problem for the RD48 with its 35bhp, but were still more of a handicap than a benefit. It has to be remembered, though, that these were Yamaha's first ventures outside of the stilted world of volcano racing, so much of the craziness should be forgiven.

Back home in Japan, Yamaha were making heavy weather, riding out a recession that had hit the country at the start of the 1960s. Original plans to return to Europe for 1962 had to be shelved, despite the fact that the RD48 had undergone a significant update and was ready to race. Ernst Degner's defection from East Germany and MZ to the West and Suzuki, seems to have borne fruits for Yamaha as well. The 1962 RD48 had an additional third transfer port in the rear wall of the cylinder, resulting in an increase of 7bhp to a claimed 42bhp at 10,000rpm. Girling shock absorbers had also been purchased during the 1961 trip to Europe and graced the otherwise unchanged chassis of the bike. The bike only had a single competitive outing in 1962, at the Singapore GP.

After that race the RD48 was discarded as having served its purpose as a prototype racer, used to establish the exact design parameters required to win 250 GPs. Yamaha had built up a consolidated base of rotary valve two-stroke knowledge. It was time to build the bike that would seriously challenge the GP hegemony that was Honda's. The stage was set for some of the bloodiest internecine fighting that the GP world had seen since the Italian factories fought each other to the verge of bankruptcy back in the 1950s.

Masuko gives the RD48 a shakedown on Yamaha's first tiny test track close to the factory at Hamamatsu.

2 Works Glory
Factory Racers 1962–8

The early 1960s were days of unbounded glory for Honda. Their first competitive machines took wins in the 125 and 250 classes in 1960. A year later they had world titles in both these classes, adding the 350 to their collection in 1962. It seemed very possible that they would be joining the record books in a couple of years, challenging MV's triumph of a hat-trick of clean sweeps of the solo titles from 1958 to 1960. Their attempt on the newly introduced 50cc class had been a mild misjudgement, resulting only in a third place in the final championship positions. The single was hopelessly outpowered by its competition, and the factory decided to withdraw for a year to prepare a suitable response to the tiny two-strokes. Honda were confident that they would take home another three titles in 1963, only the 500 as yet beyond their grasp. In truth, the easy titles were behind them. Most of the titles to come were only gained with blood, sweat, tears and a stack of money.

RD56

Honda seemed to be so busy looking ahead, they scarcely noticed what was coming up behind them, and these were Japanese two-strokes. Degner's participation in both the design and racing team, turned the corner for Suzuki and brought them the 1962 50cc title. Building on this success, they intended mounting a major challenge to Honda in the

Yamaha's first truly competitive GP-class racing bike – the RD56.

125 class for 1963. Honda made no move to re-design their 125 twin the RC142, which had been serving them faithfully since 1960. At the Japanese GP of 1962 at Suzuka, Yamaha wheeled out their replacement for the RD48, the RD56, and promptly gave Honda a good run for their money for second place. Jim Redman won easily, but Ito and Tommy Robb swapped places throughout the race, which finally went to Robb by half a second at the flag. Yamaha had a competitive 250 and Honda chose to ignore them at their peril.

The RD56 engine was little different from the early 1962 RD48, with disc valve induction to the three-port cylinder and a six-speed transmission. Carburation was via

34mm Mikunis with remote float chambers. Although there was still an oil pump giving pressured lubrication to the crankshaft, the RD56 used a 25:1 petroil mix and returned 25.4mpg (11l/100km) at race speeds. The volume of the new exhaust pipes indicated some patient dyno work that had delivered another 1,000rpm and another 5bhp. The twin produced 47bhp at 11,000rpm.

Although the engine had been given some attention, the main improvement of the twin lay in the adoption of a Norton Featherbed clone for the chassis. True to the tradition of the day, Yamaha had established the deficiency of their own chassis, looked around for the best alternative and found it in the Featherbed design. The twin-loop cradle swept down from the top of the headstock under the front of the engine, up round the back of the engine, passing round the front down tubes to the bottom of the headstock, which was extensively gusseted. Ceriani-style front forks with internal springs joined the Girlings on the rear end to bring the suspension up to contemporary state-of-the-art. There was also a friction steering damper attached to the headstock to help calm the front end. Although not adjustable, a whole range of dampers were available that could be quickly changed as required. The black watermelon tank of the RD48 made way for the orange colour Yamaha chose as their racing colour up to 1964. As well as having a new colour, the tank was more ergonomically designed, being slab-sided and giving the rider more control of the bike with his knees.

If Yamaha's finances had been in a better state, they would have won the 1963 250 World Championship. The bikes were faster than the semi-obsolete Honda fours and with a couple of GPs to sort out any bugs, they would have the reliability to go with it. As it was, the financial situation restricted them to the Isle of Man, Assen and Spa, handing the year's title to Honda on a plate, but not without giving Honda a clear warning of their race-winning potential. Now fitted with a seven-speed gearbox, their debut on the Isle of Man was quite phenomenal. Tony Godfrey, drafted in to help Ito, Hasegawa and Sunako, was recorded passing the Highlander speed trap at 141mph (225km/h), a speed only bettered by the MV and Gilera 500s and the 350 Honda. During the race, Ito led the first two laps of the 37-mile circuit, only to lose the lead during a 50-second pit stop for fuel. At the flag he was second to Jim Redman by a meagre 27 seconds and was the talking point of the Island. Hasegawa was fourth and Tony Godfrey crashed heavily on the second lap, suffering severe head injuries that ended his racing career.

At Assen, one of Redman's favourite circuits, Ito again came second to the Honda team captain, who pulled an early lead of 30 seconds and held it to the flag. Redman was out of the Belgian GP at Spa the following week, after breaking his collar bone when taking a fall during the 125 at Assen. It is debatable whether the result would have been any different if he had been able to participate in the race. Yamaha were in top form. They had fixed the fuel frothing problems they had been experiencing, despite rubber mounts for the remote float chamber. Hasegawa fell on the first lap but Sunako and Ito drew away from Taveri on the Honda four and Provini on the Morini. Fittingly it was Fumio Ito who crossed the line to claim Yamaha's first GP victory on a bike that was light years ahead of the YD Ito had thrashed through the dirt of Catalina Island five years before. Yoshikazu Sunako emphasized Yamaha's superiority by keeping the Championship challenging Morini in third place. Thirty years later with more than 350 GP wins to their name, Yamaha has become the most successful motorcycle racing factory in the history of the sport. But few of those victories could have tasted so

Works Glory

Ito leads Sunako on the short run from La Source to the line to bring Yamaha its first GP win at the 1963 Belgian GP.

sweet as the one accorded to Hiroshi Naito and his team at Spa on that Sunday in July 1963.

Despite their success, the team had to pack and return to Japan. There was no money available to extend their GP tour to the remaining European rounds. Last GP of the year was to be the first race at the new Suzuka circuit and Yamaha invited Phil Read to come to Japan and show them what he could do with the RD56. Yamaha were looking for a top GP rider to head their team in 1964, as Ito had expressed a desire to retire from racing. Read had just completed a frustrating year riding the 'works' Gileras that had been dusted off after five years in moth balls at the factory racing department. He jumped at the opportunity to ride the 250 and was leading Redman by a couple of seconds until a misfire on one cylinder with two laps to go dropped him to third place behind Ito. Redman's win gave him the 1963 title from fourth place man Provini on the Morini.

Cautious in the extreme, Yamaha were

The 1964 version of the RD56 underent detail changes based on the experiences of the short 1963 season and also received the red and white livery that was to grace all Yamaha racers for twenty years.

initially planning to run in just five of the nine European GPs, but Read convinced them to provide him with the bikes and mechanics to make it a full season. It turned out to be the closest 250 championship since the epic Mondial and MV battles of the mid-1950s. Starting the season with the 1963 orange-tank RD56, Read pulled out of the American GP at Daytona, watched by all of 200 spectators, with oiled plugs. Third place at Barcelona behind Provini (now on the Benelli Four) and Redman was more like it, but the carburation was still not right and the old chassis, with the engine high in the frame, was a marginal handler. On to the tortuous French circuit of Clermont-Ferrand, which was not expected to favour the twitchy Yamaha. With the carburation finally sorted and after removal of the steering damper, Read blitzed the Hondas during practice, recording a lap two seconds faster than Redman's best. The first half of the race was a titanic battle between the two men, Redman only managing to squeeze past Read and start building a decent lead on lap seven. Joy turned to despair in the Honda camp as Redman's machine lost its sparks on two cylinders and Read took the win almost two minutes ahead of Taveri on another Honda four.

The new RD56 was waiting for Read when he returned to England and preparation for the Isle of Man. Easily distinguished by its new red and white colour scheme, supplementary springs securing the exhaust pipes as well as threaded collars and a different clutch housing, changes inside the engine had assured Yamaha of the 1964 title. The single transfer port that had been located at the rear of the cylinder was widened and a bridge added to provide adequate support for piston and ring. This port was directed upwards at 30 degrees to the vertical, directing charge to the cylinder head and gaping exhaust port. Although the engine would undoubtedly have over-scavenged, despite the other transfers being flat, it did bring the power level to over 50bhp at 11,000rpm. This, coupled to the 22lb (10kg) weight advantage of the Yamaha, was the final nail in Honda's coffin.

RA97

Initially, Yamaha's advantage was not apparent, for the increase of power had narrowed the power band and called for different carburation from the combination that had been so successful in France. Persistent oiled plugs preceded a seizure at the Island. Saturday, 27 June 1964 must be considered as the finest day's work in Jim Redman's illustrious career. Assen was sweltering in the high eighties temperatures as so often seemed to happen in the 1960s. Redman rode in the 125, 250 and 350 races and won them all, taking the 250 race win by half a wheel from an exhausted Read. The average race speed was raised by 3mph (5km/h). Read's exhaustion was due to an equally close race with Redman earlier on the day in the 125 class. Yamaha turned up with a miniature RD56 for him to race, giving it the code RA97. It had taken the company a long time to accept that a single was not going to be competitive in the fierce battles that had been developing between Honda and Suzuki. The RA97 used the RD56 frame and had a bore and stroke of 44 × 41mm, capacity of 124.6cc and an eight-speed gearbox. It was claimed to produce 28bhp at 13,000rpm, close to the output of the four-cylinder machines Honda were running. Redman won the 125 race from Read by 6 seconds, but for most of the race they had been at it hammer and tongs and the RA97 was clearly competitive. It was a remarkable debut race, but the RA went back to Japan and was never seen again in Europe.

On to Spa and Yamaha, aware of the World Title hanging in the balance, had hired Mike Duff to join Read in the team and guard his back whenever possible. Read's engine seized while he was going flat out down the Masta straight at the back of the circuit and he almost took out Redman and Duff who were tucked in behind him. Redman's 140mph (225km/h) trip through the shrubbery shook him up enough for Mike Duff to steam into a commanding lead which he held to the flag for his first GP win. With the bikes finally sorted, the rest of the season turned into a Yamaha benefit with wins for Read at Solitude, Sachsenring, Rouen and Monza. Honda had panicked into a crash programme to develop their six-cylinder 250 planned for the 1965 season. Making its debut at Monza, it was fast but fragile, losing a comfortable lead due to a bad misfire on full throttle. Read's Monza win brought him his first rider's title, Yamaha their first manufacturer's title and the two-stroke its first 250 title.

CONCEPTION OF THE V-4S

The appearance of Honda's six-cylinder 250 at Monza stunned Yamaha, just as the machine's runaway 1–2 at the Japanese GP five weeks later prompted an emergency meeting to consider the correct strategy to adopt to meet this challenge. There was agreement that the RD56 was approaching its maximum development form, only water-cooling and the contemporary practice of raising primary and secondary compression being options that might keep the twin competitive through 1966. Two-stroke tuning along these lines in the 1960s inevitably led to narrower power bands and extra gears to keep the engine on the boil. Yamaha rightly perceived the limitations that this evolutionary path would encounter. Instead they opted for revolution rather than evolution. They would take a leaf out of Honda's book and take the path of lowering reciprocating

A worm's eye view of the engine of the 1965 RD56, that was campaigned by a number of the most favoured riders in different European countries until 1968.

mass, raising engine speed and produce the world's first V-4 250 two-stroke. The obvious starting point for the design was the RA97 125 twin that had began so well in Assen. Using the 125 as a testbed for their ideas, they would look for introduction of the new 250 halfway through the 1965 season. The RD56 would have to perform a 'damage limitation' exercise until the V-4 was ready, and consequently would undergo its last development phase prior to the start of the race season in April.

The winter work on the RD56 managed to squeeze a further 3 or 4bhp out of the engine to bring it up to a total of 55bhp at 11,500rpm. In comparison, the square four Suzuki RZ65, which had been unsuccessfully campaigned since its debut in 1963, was claimed to produce 56bhp at 12,850rpm by 1965. The first versions of the six-cylinder Honda 250, the RC165 were giving 56bhp at 17,000rpm, so there was not much between them as far as power was concerned. Despite the mechanical complexity and the valve-train penalty of a four-stroke, the Honda was the lightest of the three machines at 248lb (112kg) The water cooling of the

Mike Duff takes charge of his RD56 for the second and last US GP of the 1960s at Daytona in March 1965. He finished second to Phil Read.

Suzuki gave it a 44lb (20kg) penalty with the RD56 positioned midway between the two.

So the 1965 season got off to a start at the second and last USA GP of the 1960s, and Yamaha were presented with an easy win when Redman, principled man that he was, refused to start after a dispute with the organizers about start money. Read first, Duff second. At the German GP, Redman crashed during the 350 and dislocated his shoulder, excluding him from the 250 race. Read first, Duff second. A week later at Barcelona, Redman did not feel fit enough to race. Read first, Duff third. The French GP at Rouen saw Redman streak into the lead only to retire at half distance with gearbox problems. Read first, Duff did not finish. At the Isle of Man, Read set the first 100mph (160km/h) lap on a 250 from a standing start, but the engine seized on the second lap. Mike Duff took second place. At the Dutch TT, Read was in superb form, streaking away from Redman to win the race with Duff finishing third. It was now all over bar the shouting, despite wins by Redman in the Belgian and the East German GPs. The Read–Duff 1–2s in Czechoslovakia and

Works Glory

Phil Read lapping a Honda privateer on his way to victory at the 1965 US GP.

Ulster clinched the title for the second time and the simple RD56 twin had proved the equal of Honda's complex technical marvel. Nineteen sixty-five was the last year that the RD56 was run at GP level but it enjoyed post-retirement glory in the hands of top national riders in France, Holland and Britain, winning national races through to the end of 1968, when the supply of spares was finally exhausted. It earned its place in history as the bike that was born from the development of Yamaha's understanding of the powers at work in the two-stroke engine. Compared to many of its contemporaries, it was a well-balanced machine in its final form, handling satisfactorily, producing good power, and able to withstand the over-revving that Duff and Read claimed they frequently inflicted during 1965, without failure. As this elegant classic of well-proportioned racing bike slipped into the shadows, the spectre of a behemoth moved into the spotlight. The RD05 V-4 250 Yamaha, the most complex motorcycle Yamaha were to build for nearly 20 years, had drawn its first breath and screamed its birth to the wind.

RD05 – THE FIRST V-4

The rolling testbed which Yamaha had decided to use as a prototype for the RD05, the water-cooled RA97, had appeared for practice on the 1965 Isle of Man, set third fastest time in practice and won the race by just 8 seconds from Taveri on the four-cylinder Honda. It had been a close thing as a piston ring broke at Creg-ny-baa, just 3

The 1965 RA97 125 was Yamaha's first water-cooled racer and served as the testbed for the mighty V-4s to appear later that year.

miles from the finish. Mike Duff's machine had lost a piston ring at Kate's Cottage, just half a mile before the Creg, and it had cost him second place. The RA97 ridden by Duff and Read was simply a water-cooled version of the bike Read had ridden a year earlier into second place at the Dutch TT. Bill Ivy had been given the air-cooled version for the Isle of Man TT and had managed a good seventh place, despite stopping to change plugs half-way round the last lap. Yamaha had wisely chosen to fit a water-pump to the new RA97, rather than rely on the thermo-syphonic versions that were common on most other contemporary two-strokes, including the 'Whispering Death', the Suzuki RZ65 square-four 250. A touch over 30bhp was claimed for the engine at 13,000rpm. It looked like being another Read victory at the Dutch TT ten days later, but with a 20 second lead at the start of the last lap, a small-end broke and his race was over. Duff got the best of a three-way duel with the Suzukis of Katayama and Andersson, but a post-race stripdown revealed that it would not have survived another lap, with both big and small end bearings breaking up. Completely unfazed by two wins from two races, Yamaha felt the two races had enabled them to assimilate enough information to develop a reliable machine for the 1966 season, and they were flown back to Japan.

In the meantime, the first version of the RD05 was ready. Yamaha still had no decent test circuit of their own, and rather than test it at Honda-owned Suzuka it was flown over to England for a secret test session at Snetterton. Most people had expected it to be a square four along the lines of the RZ65 white elephant Suzuki had spent two years trying to make competitive. It is unclear if the decision to use a V configuration was made to avoid the charge of plagiarism, or if a genuine advantage was seen over the back-to-back format of the Suzuki. The chas-

Mike Duff (9) and Bill Ivy (24) on RA97s sandwich Hugh Andersson and Yoshi Katayama on Suzuki RT65s at the 1965 Dutch GP. Duff won with Katayama second.

A jet-lagged Bill Ivy, drafted in to replace an injured Mike Duff, gets the lowdown from Phil Read on riding the RD56 at the 1965 Japanese GP at Suzuka.

sis of the RZ had a wheelbase of 56in (1,400mm), which would have been considered long for a 500. The V-4 engine would not necessitate a long wheelbase, but it would be an engine with a high centre of gravity.

The engine dimensions of 44 × 40.5mm gave a strong indication of the design philosophy; couple two RA97 engines, already good for more than 30bhp, to a common

Works Glory

The mighty V-4 250, the RD05, as it first appeared during 1966. The high rear swing-arm pivot point, clutch and lack of magnesiun components, identify this as a 1966 model.

crankcase. The front cylinder pair was mounted horizontally, the other pair leaning forward at 20 degrees to the vertical, to form a 70-degree V-4 engine. The two separate contra-rotating crankshafts delivered their power to the gearbox via a jackshaft located between them. The jackshaft also drove the two magnetos providing the sparks and idler gear which in turn drove the water and oil pumps, both mounted behind the upper cylinders. True to tradition, Yamaha did not want to rely solely on a petroil mix to lubricate the engine and a total of seven oil lines ran off the oil pump to pressure lubricate the main bearings and crankshaft big ends. The eight-speed gearbox was driven via a dry clutch of enormous diameter on the right-hand side of the engine, itself driven by the primary gear on the jackshaft. Carburettors of 25mm throat diameter were used to provide the charge via the disc valves on each crankshaft end. These differed from earlier models in that they used an integral float chamber, rather than the rubber-mounted remote chamber first seen on the YDS1R seven years before. Exhausts on the front cylinders swept under the engine, whilst those on the upper cylinders passed back over the engine inside the tubes of the cradle frame and outside of the twin vertical shocks at the rear of the bike. Alloy heat shields prevented the riders roasting their legs.

The RD05 was water-cooled. Initially both water-cooled and air-cooled versions were constructed, the air-cooled version offering a 53lb (24kg) weight advantage. Knowing that the bike was at the start of its development cycle, and worried about the difficulty of cooling the horizontal cylinders effectively,

Works Glory

the factory rejected the air-cooled version. Dyno tests had also shown the air-cooled version to start losing power much earlier than its sibling, and although power was never to be a problem with the V-4, it was perceived as such. The initial version of the RD05 produced approximately 65bhp at 13,500rpm. This monster of an engine was shoe-horned into the same frame that the RD56 had used, which was now supposed to cope with a 20 per cent increase in power. It could not and this more than anything compromised the effectiveness of the RD05 during its first season. The radiator needed to be placed high on the front down tubes to avoid fouling the forward cylinders and added to the high centre of gravity. Mike Duff tells tales of 33lb (15kg) lead weights being clamped to the frame rails under the engine in an attempt to compensate for the top-heavy feel of the bike, with some measure of success. Realizing that the 320lb (145kg) machine would be tough to wrestle down from its 150mph (240km/h) top speed, a massive double twin-leading shoe front brake was fitted and the alloy drum almost completely filled the front wheel.

The RD05 was not an attractive machine and had none of the elegance of the RC166 that it was to spend its life fighting on the race tracks. Initially the six-cylinder Honda was a camel to ride, but under Mike Hailwood's guidance, it and in particular its big brother, the RC174 350, was to become one of the best balanced machines that any of the Japanese companies built in the 1960s. This could never be said of the RD05. The riders built up a working relationship with the machine that eventually enabled them to win GPs and world titles. But it was always an uneasy relationship, for the bike always seemed to be on the point of biting back.

The first time it bit back was at its debut race at Monza in 1965. Yamaha's decision to give the bike its debut so late in the season was puzzling. The title had already been won with the RD56, and there was no need to enter the machine at GP level prematurely. Perhaps it was a wish to get back at Honda after the shock appearance of the RC166 a year before. Anyway, it was a wrong decision, the bike not seeming to take to the uncharacteristically cold and wet Italian weather. Read had some problems getting the Yamaha to show any signs of life at all at the start. Finally a couple of cylinders woke up, but it took another half a minute before all four were running. The slick surface of the Monza track and the poor handling of the bike did nothing to inspire Read with the confidence necessary to fight back into contention. Half-

The story of the 1967 250 GPs. Mike Hailwood battled the Yamaha riders all year to take the title by the closest of margins. He won the war, but this battle at Clermont-Ferrand he lost to Bill Ivy and Phil Read.

Works Glory

way through the race, the front cylinders went back to bed, their sparks being drowned in the increasingly heavy rain. After a stop in the pits to change plugs, Read rejoined the race, but finished a poor seventh. During practice for the end-of-season Japanese GP, the RD05 bit again; Phil Read's engine seized at high speed and Mike Duff dropped his on a fast corner. Duff badly broke his leg, causing him to miss much of the 1966 season and Read was too bruised to ride in the race. A panic call to Bill Ivy back in the UK got him over to Japan in time to finish third in the 250 race despite stopping to change oiled plugs. Ivy went back home with a works contract in his pocket for 1966 and the famous partnership of Read and Ivy was a fact.

Despite the feedback Read had provided the factory on the deficiencies of the V-4, little change was made to the machine for the 1966 season. To make matters worse, the competition was now the almost invincible combination of Hailwood and Honda. The RC166 that Redman had been riding was itself of marginal effectiveness. This was made crystal clear by Mike Hailwood after the 1965 Japanese GP, when he replied to a journalist's question of how good the bike was with the succinct 'Bloody awful'. Honda got to work and did improve the chassis significantly by the start of the 1966 season. It turned out to be an absolute walkover for Hailwood. He won every 250 race he finished, a total of ten of the twelve GPs. Honda boycotted the Japanese GP, miffed at its move from their own track to Fisco, and

Bill Ivy at Ginger Hall on the Island, on the way to the third of his four victories on the RA97 during 1966.

Hailwood was unable to compete in the 250 Ulster as he would have exceeded the maximum race distance permitted by the FIM on a single race day after his 350 and 500 rides. Read managed four second places to Hailwood mid-season at Assen, Spa, Sachsenring and Brno, but the superior speed of the machine did not compensate for the awful handling.

RA31 – THE 125CC V-4

Things were going slightly better in the 125 class, where Bill Ivy, on the water-cooled RA97 twin, was challenging the new five-cylinder 125 Honda ridden by Luigi Taveri. Ivy won three GPs to Taveri's five, and he managed consistent finishes in the ones he did not win. Taveri ended up taking the title by four points from Ivy. Despite the comparative failure of the 250 V-4 to deliver the goods, Yamaha remained faithful to the concept and decided it should be applied to the 125 class as well. By the Ulster GP in September 1966, a mini V-4, confusingly encoded the RA31 was available for use during practice. Intended for Bill Ivy, it was Phil Read who did a few laps of the Dundrod circuit before a main bearing broke and the bike went back in the van. Bill Ivy was a non-starter after a bad concussion from a fall at a British national meeting. In all, 1966 was a year Yamaha would rather forget. Honda were virtually invincible taking the marque title in all the solo classes, winning all four solo races at the Czechoslovakian GP, and winning the rider's title in the 125, 250 and 350 classes. Yamaha were determined to mount an effective challenge to them in 1967, using the 125 and 250 V-4s.

The 125 was a miniature replica of the 250, using the same frame and general design. In fact when lined up together, only the smaller front brake and absence of a frame mounting plate at the rear of the engine could distinguish it from its big brother. With engine dimensions of 35 × 32.4mm, it needed an extra speed in the gearbox to keep it in the powerband from 15,500rpm to 17,000rpm. The oil pump had a total of twelve lines that ran to the main bearings, rotary valves and big ends. The carburettors were tiny magnesium 22mm Mikuni Monoblocs. Almost the whole engine was magnesium. The 125 had benefited from the lesson learned during the 1966 season with the 250. It had been too heavy, and a weight-reduction exercise resulted in a smaller, compacter engine, clearly distinguished by the smaller clutch and crankcases. In addition magnesium was used for the crankcases and carburettors. A few extra bhp were squeezed out of the 250, with the use of 27mm carburettors, bringing its total maximum output up to 70bhp, and with the weight down to 130kg, it should have had a noticeable speed advantage over the Honda. The 125 tipped the scales at a respectable 232lb (105kg), surely enough to make it a

The 1968 RA31 revealed. Despite the legend that grew around the bike, the V-4s were not difficult machines to work on.

Honda 'killer' on the GP circuits.

The only problem was that Honda had decided not to compete in the 50 and 125 classes. The cost of racing all the solo motorcycle classes, plus a Formula 1 car team was proving prohibitively high. With their intention of making a final supreme effort to gain the 500cc class rider crown, the one title that had eluded them until now, they sacrificed the support for the legendary 125cc five-cylinder and 50cc twins. It must have left Yamaha with a bad taste in their mouth, but they decided if it could not beat both Hondas and Suzukis, it would just have to beat Suzukis instead. And this they did, in ten of the twelve GPs of 1967, eight going to Ivy and two to Read. The Suzukis won at Hockenheim, both RAs having been brought down when Francesco Villa dropped his Montesa while being lapped. Only at Imatra in Finland did the RA really bite back when Ivy had to stop twice to change oiled plugs and ended up second to Stuart Graham on the Suzuki twin. Suzuki were so impressed by the performance of the RA31, that they built themselves a copy that was raced by Graham into second place behind Ivy at the Japanese GP. Three months later Suzuki announced their withdrawal from GP racing and after a few Far East races during 1968, the new Suzuki was banished to the corner of a dark and dusty storeroom near the race shop.

THE IVY/READ/HAILWOOD BATTLES

If only it had been so easy for Yamaha in the 250 class, but once again the legendary combination of Hailwood and Honda deprived them of the riders title. This time it was close, so close that it could even be argued that it was really Read's title as he ended the season with more points than Hailwood. At the time, the seven best results of each

The last version of the RA31, most easily distinguished from its big brother by the smaller front brake and carburettors.

The last version of the RD05, with a lashed-up electronic ignition replacing the racing magnetos run for most of the year. The left-hand gearchange suggests this is neither Ivy's nor Read's bike.

rider were taken into consideration, giving both riders fifty points. The English language version of the FIM regulations stated that in the event of tied points, an eighth result would be counted, a second place for Read and a third for Hailwood giving Read a two-point advantage. Read was busy celebrating his third World Title, when it was discovered that the French version of the regulations stated that the number of wins should be counted in the event of a tie. Hailwood had five to Read's four, and he was declared World Champion. The French copy of the regulations had been updated in 1964, but the FIM had 'not yet been able to issue an updated English translation'.

The RD05 had almost become reliable, with Read finishing eight of the thirteen GPs. One of the tricks that the mechanics had discovered, was the need to run the vertical cylinders slightly richer than the horizontal cylinders. The vertical cylinders were cooled by water that had already been pumped through the horizontal cylinders and was thus slightly warmer. The height of the float in the 27mm carburettors could be externally adjusted and a small inspection glass was provided on the side of the float chamber to assist the adjustment. Initial problems that the RA31 had experienced with broken piston rings had been solved by the use of a very thin 0.8mm ring. The transmission bearings were also provided with pressurized lubrication via the hollow main and lay shafts. Transistorized ignition was used on the RD05 for the start of the 1968 season, but both riders switched back to the tried and tested magnetos after a couple of races. Responding to the continued complaint about the handling of the 250, the frame was provided with adjustable trail and rake. This was achieved through the use of steel shims of different thickness to pack a pivoted lug where the steering column was bolted to the lower triple clamp. It helped but the handling of the RD05 was never properly sorted.

Once again Yamaha were denied the

Works Glory

The 1968 needle match. Bill 'let' Phil win the East German GP at the Sachsenring, and must have bitterly regretted it a few races later.

chance to redress the defeat Honda had inflicted on them in the 1967 season. Honda announced their total withdrawal from GP racing at the end of the 1967 season, and the Yamaha team riders were left to fight each other, no other machines in either the 125 or 250 classes even approaching the same performance league. The factory had decided to allow Read to win the 125 title and Ivy the 250, but the infamous needle match developed which resulted in Read 'stealing' Ivy's 250 title, once he'd been 'handed' the 125 title by Ivy. With no one else to race, it seemed inevitable that these highly competitive riders would end up racing each other. It is only sad that it resulted in so much bad blood and that Ivy, killed in July 1969 riding a Jawa during practice for the East German GP, was never able to accurately reflect his enormous racing talent in more than the single 125 title of 1967.

Yamaha withdrew from direct factory-supported GP racing at the end of 1968. With the FIM regulations restricting 125 and 250 machines to two cylinders and six gears, looming in the distance, the V-4s were technological masterpieces, instantly declared obsolete. Both machines have become legends in the annals of motorcycle racing history. They set standards of performance that were to take engineers working within the FIM's restrictions close to 20 years to surpass. Even more important than the level of performance these machines achieved, was the insight into the incomprehensible world of the two-stroke engine, given to Yamaha engineers through their development. The design of Yamaha's production racers was directly influenced by work done on the V-4s. They were Yamaha's contribution to the legacy of the golden age of Japanese racing, and it was a legacy Yamaha could be proud of.

3 Growing Pains
TD1 Production Racers 1962–8

THE ASAMA RACER

When Taneharu Noguchi pushed his YDS1R into a briefly held third place in the 1959 Asama race, it represented the beginnings of the reputation Yamaha 250 production racers were to hold far into the 1980s. Giant-killers, humblers of factory machines and riders, machines to be feared and loathed by the competition for their sheer effectiveness. Sadly Noguchi did not complete the race, his broken transmission testimony to the long road facing Yamaha before they would achieve the technical excellence for which they strived. With the YDS1R, their journey of enlightenment began; with the TD1C they were no longer travelling into the unknown.

The YDS1R was a YDS1 street bike, half of which had been replaced by a separately purchased race kit. The kit consisted of cylinders, heads, pistons, magneto, carburettors, exhausts, close-ratio gears, tank, seat, clip-ons and tachometer. For those wishing to tackle circuits like Asama or Catalina, there were scrambler exhausts passing up over the engine side-covers with heat guards to prevent the rider's legs being burnt. The cylinders were alloy with iron liners, offering a significant weight saving over the cast iron cylinders of the YDS1 street bike. More radical port timing was used on these cylinders and it was this more than anything else that compromised the reliability of the 1959 version. The poor thermodynamic bond between iron liner and aluminium barrel meant that the cylinder wall around the exhaust was not cooled well. The aluminium pistons expanded too much and seizures were common. The exhausts were faithful copies of those fitted to the German Adler RS250 and did little to help scavenge the engine effectively. If anything,

An historic moment as Noguchi squeezes his YDS1R past Suzuki's four-cylinder Honda RC160 at the 1959 Asama race.

37

Growing Pains

they contributed to the unreliability by not extracting the exhaust gases at high engine speeds resulting in detonation and piston failure. The YDS1R marked the first appearance of the 27mm Japanese Amal 276 carburettors with remote float chambers intended to prevent frothing. These were to live on through the generations of TD1s to follow, but their efficacy was questionable from the start. Perhaps the remote float chambers reduced frothing, but the acceleration and deceleration of the bike forced fuel to surge or ebb. Not being an especially strong engine, the YDS1R did not suffer too much from this problem but by the TD1C it was the major cause of TD1 engine seizures.

It is odds on that the failure of Noguchi's YDS1R at Asama was due to the clutch mounted on the end of the crankshaft. As it spun at engine speed, it was very easy to damage during gear changes; the damage was usually terminal, most often involving a snapped crankshaft, occasionally transmission mainshaft failure. In extreme cases the outer clutch drum would explode. The clutch was especially vulnerable, as the 56 × 50mm race engine would rev to 8,500rpm, 1,000rpm higher than its street-based brother.

One other characteristic of the Asama racer was so bizarre as to defy explanation; its colour scheme. It 'featured' a reddish brown tank with Yamaha in small white letters, white front fork shrouds and a pink seat. The pink seats were the first to go. Some of these machines found their way out of Japan and into the Australian and

Yamaha sent Sonny Angel this YDS1R, and a box of assorted spares with which to compete in the 1960 250 race at the Isle of Man. The spares were long gone before he even reached the Island.

American markets. In America, Larry Beale was contracted by the new Yamaha organization to race the bikes at Dodge City, but inevitably his machine suffered a seizure and he retired. Reports returning from Australia confirmed that there were fundamental flaws in the engine materials. Yamaha decided that there was a need to update the kits for 1960, or else their policy of exposure through racing success was doomed to failure. The seizure problem was tackled by the casting of pistons with higher silicon content coupled to the replacement of the iron liners by an anodized all-aluminium cylinder. Another improvement was the switch to a single 1.5mm ring in place of the dual 2mm rings previously used. This killed two birds with one stone. It was a precaution against ring flutter that would occur at lower revs with a 2mm ring and enabled a reduction in friction by having the single ring.

With these major engine improvements and a half-hearted attempt at beefing up the clutch, Yamaha continued their shot at domination of the Australian and American 250 classes. In Australia, Kel Carruthers was sweeping away all opposition on a four-cylinder Honda RC162, but Asama racers in the hands of Ken Rumble and Mick Dillon were often at the head of the bunch battling for second place. In the US some clubmen bought themselves race kitted YDS1s and did battle with the Duactis, Motobis and NSUs dominating the class at the time. For Europeans, the most famous of these brave souls was Sonny Angel, who was so impressed with the speed of his YDS1R, that he decided to do some 'evangelical' work in Europe and brought his bike over to the UK for the Isle of Man TT and a few other meetings. The trip was a total disaster. First time out for practice at Brands Hatch the engine seized in a big way. During a race at Silverstone before the TT, he was lapped on the fourth lap by Mike Hailwood on a Ducati. During practice for the TT, he was unable to qualify, suffering persistent seizures and holed pistons. Sonny Angel recalls his trip to Europe as a frustrating experience.

> When I arrived on the Isle of Man in June 1960, the factory Yamaha 250 had no pistons as all originals and spares had seized in testing and racing at Brands Hatch and Silverstone.
>
> Hepolite supplied two sets of pistons for an MV scooter which I had to modify by relocating the ring locating pegs. I also had to machine the cylinder head combustion to 10:1.
>
> It ran well and revved 8,400/8,600 on the highest gear ratio I had available giving a top speed of about 116mph. On the third lap of practice, the locating peg came out and broke the lower skirt of the piston at the Quarry Bends just past Ballaugh Bridge.
>
> Back in California we ran later technology pistons and ran it with some success in California road races during the early 60s.

THE TD1 IS BORN

The race kits were to remain on sale for the 1961 and 1962 seasons as Yamaha had their hands full with their GP programme that took them to Europe in 1961. Realizing they had a lot of work to do to challenge the existing order in 250 GP racing, the factory decided to use 1962 as a sabbatical year to prepare a better racer for a repeat visit to European GPs in 1963 and to implement a major update to their clubman racer programme. This resulted in a new machine appearing from the doors of the Numatu factory, based on the newly introduced YDS2 street 250, but already fully race kitted and ready to go. The bike bore the name TD1 and was the first of more than 500 production racers that were to totally dominate the clubman's sport by the end of the 1960s.

The origins and exact specification of the

Growing Pains

The official photograph of the 1962 TD1. The heavy photographic re-touching plainly visible only increases the mystery surrounding the first of the TDs.

The Tokyo Motorcycle show of November 1962 featured a fully race-kitted TD1 on the Yamaha stand. The rear end of a YDS1R is visible to the right of the TD.

Growing Pains

A TD1A or early TD1B? Very difficult to distinguish between them, but the shallow taper on the exhaust suggests a TD1A.

Bill Ivy on a TD1A chases Phil Read on the RD56 at Brands Hatch in May 1964. A portent of things to come.

TD1 have all but vanished in the passage of time. There are some who claim that the TD1 never really existed as a model separate from the TD1A, which was exported in large numbers to Australia and the USA, even reaching the UK in minute numbers. Yamaha themselves say the the TD1 was actually a 'YDS2R' or YDS2 with the tank, race seat and racing front end fitted, and there are photographs to support this. It is also clear that the pukka TD1 delivered to the USA and Europe differed at the very least in exhaust pipe. The first models to be tested by *Cycle World* had short bulbous RD48-type exhausts fitted, as did at least one machine that arrived in the UK at the

41

end of 1962. TD1As seen later in 1963 have longer, tapered, reverse cone exhausts, somewhat more compatible with the engine characteristics. Whether these differences reflect a different model is doubtful. Most likely they are simply the first examples of Yamaha's willingness to use anything that came to hand to get the machines out the door at the start of the US racing season, in time for homologation. This was to happen again with the TD1B and even in the 1970s with the TZ750.

The YDS2 was only marginally different from the YDS1, so the TD1A did not differ so very much from the YDS1R. The frame was of the same basic design as the street machine, but used lighter gauge steel to save 4.5lb (2kg). The same swing arm, front forks and rear dampers were used, but the suspension at both ends used far stiffer springs. The front wheel was fitted with an alloy twin-leading shoe drum brake with an air-scoop cast into the backing plate. The high, six-gallon (27litre) capacity fuel tank was replaced by a more conventional slimline model that was painted a dirty orange, Yamaha's colours at the time. The racing fraternity was pleased to see a black seat re-installed.

The anodized cylinders were retained for the TD1A, although there had been some experiments with the use of chromed cylinder walls, and a couple may have been fitted to machines sold to customers. Yamaha were aware that the anodized cylinder walls did not effectively support a coherent oil film, and they rightly attributed many of the seizures of the YDS1R to the breakdown of this film. Chromed walls maintained a better oil film but the coating process was imperfect and the chrome would lift and flake off. A close-ratio gearbox moved gears two through five closer together at the expense of the gap between first and second, which was now enormous and severely hampered the TD1A on tight circuits. Carburation and ignition were identical to that of the Asama racer, but fortunately the exhaust was changed. Although a seemingly original design rather than an Adler copy, the A's exhaust was only marginally more effective. It was a little too short, meaning that it would have been more effective with an engine revving higher than the 'safe' 9,500rpm specified. More importantly, it was of too low a volume to make it an effective scavenger of waste exhaust gases.

Despite this, it received a good press in both the USA and UK. None of the quirks with which it would become associated were mentioned. The very low first gear, the sensitive clutch, the brake fade, the rigid rear suspension, the under-damped forks, all went without comment. In total about 120 TD1As were built, the vast majority going to the USA and Australia, with less than ten ending up in the UK, the only European country where they were sold. Despite their problems, they were able to win races, and Neil Keen created a historic milestone by winning the Dodge City Labor Day 250 race in September 1963. This success was not matched in the UK, despite the use of UK importer Monty Ward's TD1A by a rising star, Bill Ivy. Bill had a couple of good rides in the UK and came close to winning an international race in Barcelona on Frank Sheene's A, but neither Bill nor any of the other A riders managed a win in 1963. The opposition from two-stroke Cottons, Greeves and four-stroke Aermacchis, Ducatis and Honda twins was too strong. In August Bill finished second at Brands Hatch and John Field managed another second place at Snetterton a month later. It was to be July 1964 before Brian Warburton used a TD1A to win a race at a national meeting at Aintree. At Daytona that year, TD1As had taken third, fourth and sixth places, demonstrating their proximity to the big-time wins. Within a year TDs were to hit the big time with a vengeance.

The official portrait of the TD1B, although the earliest models were supplied with the last of the traditional dirty orange fuel tanks.

TD1B

Yamaha had spent two years selling TD1As and keeping their ears to the ground for feedback from their customers about the perceived shortcomings of the machines. Simply put, they handled like a hinged door, couldn't pull the skin off a rice pudding and were as fragile as fine-bone china. The transition from A to B involved a significant engine update, whilst the chassis marked time. Once again the production of the new racer coincided with an update to the street model to which it was so closely related. The YDS3 underwent a number of changes that seemed to be directly related to experience gained at the hands of the TD1A racers. The bottom end of the engine was up-rated with 5mm thicker crankshaft and improved small and big-end bearings. Undoubtedly this was the result of the frequent failures experienced by TD1A riders using YDS2 cranks. Rather than simply update the racer, the YDS3 source machine was updated to the general benefit of the tens of thousands of customers that bought the street 250. Racing was undoubtedly improving the breed.

With the bottom end of the engine improved, the cylinders and pistons could be updated for the racers. A step to improving the longevity of the engine was taken by switching from anodized cylinders to the porous chrome plating that had been developed for use on the works RD56 during 1964. The cylinder port timing was unchanged from the TD1A, although repositioning of the cylinder studs enabled the crankcase entrance to the transfer ports to be enlarged. Although the cylinders remained of the same dimensions, the overall engine timing was changed quite significantly by modifications performed on the piston. In order to extend the exhaust period a 2mm deep notch was cut in the exhaust side of the piston crown, effectively raising the exhaust port height, and enabling the exhaust pipe to scavenge the cylinders more effectively at high revs. On the inlet side, 9mm of the piston skirt was cut away, resulting in the inlet port opening 90 degree BTDC. This was the up side in that the engine was given more time to fill the crankcase fully before the transfers opened. The down side was that the inlet closed only 21 degrees before the transfers opened which would make the machine more difficult to start when the natural resonances of

43

the running engine were not present. Minor changes were made to the carburation and ignition, with a slightly smaller main jet and slightly later ignition timing. The cylinder changes were only effective with a completely re-designed exhaust pipe, which bore some resemblance to the pipe design that was to persist into the 1970s.

The pipe grew a lot longer and fatter. The fatter centre section of the pipe ensured that the pressure waves traversing the pipe were of sufficient magnitude to suck hard during the period between the opening of the exhaust port and the opening of the transfers and blow hard as the fresh charge from the transfers was over-scavenged and entered the pipe. Unfortunately the growth in length hampered the ability of the engine to rev by causing the stuffing wave, responsible for pushing any lost fresh charge back into the cylinder, to arrive too late. Yamaha were still learning the black art of two-stroke engine design and most of their experience had been garnered on the GP race tracks of Europe with the rotary valve RD56. It was becoming clear to them that they were very different beasts altogether, with their own special problems and solutions.

Despite the less than optimal exhaust design, the TD1B revved to 10,000rpm, a gain of 500rpm over the TD1A, and squeezed out a fraction over 30bhp at that engine speed. It was marginally faster than its predecessor, but most important of all it was more reliable. Yamaha were determined that it would be the US market that demonstrated this and they hastily cobbled together the one hundred units that were required to meet AMA homologation requirements in time for the 1965 Daytona race. In the mad rush to get the bikes assembled on time, some non-standard TD1B parts were pressed into service, resulting in some machines being sold with YDS3 transmission components and, in a few cases, pre-production sand-cast crankcases. It was all worth the effort, though, for Daytona was the first of what was to become many Yamaha clean sweeps, with Dick Mann, the doyen of American racers, narrowly beating Gary Nixon and Buddy Elmore all riding TD1Bs. It was this race more than any other that was to mark the beginning of the end of the dominance of the four-stroke in US road racing. Within the year, the TD1B had wiped up the opposition in the 250 and 350 classes. During 1966, 500 and open class races were often won by TD1Bs.

One of the most successful TD1B riders was Frank Camillieri, who designed and sold his own frame to house the TD1B engine. Journalist Kevin Cameron recalls the elation that went with Camillieri's total dominance:

> In those days you could enter your 250 in the 350, 500 and open classes in club racing. Camillieri won them all one year, and the big-bike riders went out on strike; we won't race anymore unless he is out of here. They compromised; Camillieri could ride, but would not be scored except in 250. After that, he would build up a big lead and stop out on the course, waiting for the big four-strokes to come toiling around. He would pull out after them from a standing start and catch and pass them down the straight. Or he would pull up a half-minute lead and wait playfully a few feet short of the finish line on the last lap. When the smoking bored-out Triumphs and Nortons came wobbling into view, he would grin at them and push across the line first.

Daytona 1966 was another TD1B benefit with Bob Winters heading home John Buckner and Ralph White, despite Harley-Davidson entering works Aermacchis that could do no better than fifth place. The same story could be found in Australia, but in Europe the TD1B was still struggling to defeat the European 250s of Italian, British

Reg Everett on an early TD1B with a TD1A fairing bearing the Yamaha name. Everett was the UK's most successful rider on the early TDs.

or Spanish origins. Greeves, Cottons, Aermacchis, the odd Ducati, Bultacos and Villiers made life difficult for the few TD1B riders. There were very few machines available in Europe during 1965, the UK ending up with less than ten machines and a few engines. In France, Daniel Lhéraud took his TD1B to third place in the National Championships. In the UK, Reg Everett was the most successful B owner, finishing second to Dave Simmonds on his ageing CR72 Honda in the British Championship. Don Padgett started campaigning the bikes, undergoing an apprenticeship that was to come to fruition with the arrival of the TD1C.

Twenty more machines trickled into the UK during 1966, and both Holland and Germany also saw the first arrival of very small numbers of the TD1B on their circuits that year. The 1966 model had a number of minor changes. The old watermelon orange tank had been replaced by a pointed-nose white tank with gold Yamaha lettering on the side; very stylish. The front suspension featured stronger springs. The remote carburettor floatbowls were no longer bolted to a lug on the crankcase, but to a strut suspended from the main upper frame tube. The exhausts were no longer bolted directly to the frame but to a spring-loaded bracket attached to the frame. In addition the pipes were coated with crack-resistant paint instead of the chroming that had been done before. By the end of 1966, approximately 300 TD1Bs had been sold around the world and had built the foundations for the worldwide success that its successor, the TD1C, would achieve.

TD1C – THE TD1 MATURES

Road machines were undergoing a two-yearly cycle of model update, so it came as no surprise that a new 250 YDS was announced for 1967. It was the YDS5, the '4' designation having been skipped as being unlucky, synonymous with death in Japanese. The YDS5 had a single big plus point and a single big minus point. The plus point was the movement of the clutch from the crankshaft to the mainshaft of the transmission. The minus was its replacement on

Growing Pains

the crank by an electric starter. Fortunately this totally superfluous piece of equipment was missing on the YDS5's racing brother, the TD1C. At last the rider did not have to take such care with gear changes, or constantly have to adjust the clutch after a couple of fast take-offs. This change alone would have had many TD1B owners reaching for their cheque books, but there was more good news to come. The gap between first and second gears was narrowed again to a value that was largely unchanged during the 1970s. Although the transfers and exhaust ports on the C-cylinders were unaltered, a couple of extra pseudo-transfers were carved in the wall of the aluminium cylinder.

Legend has it that these extra gulleys in the TD1 cylinder walls first appeared in the US in the 1966 season, but the name of the enterprising tuner who cut them has disappeared in the mists of time. The gulleys were 12mm wide and 3mm deep and ran parallel to the side of the inlet port then bending inwards above the inlet port to run parallel to the transfers. Corresponding to these transfers were windows cut into the skirt of the piston adjacent to the section of the gulleys running parallel to the inlet port. These extra ports added a marginal amount to the power output of the engine by assisting in the scavenging of the exhaust gases. Equally important, they assisted its reliability by washing the underside of the piston crown as well as the small end bearing with a cool slippery petroil mix. Another 500rpm were won on the top end with the improved engine and Yamaha, deciding to play safe, reduced the piston ring thickness a further 0.3mm to 1.2mm.

The exhausts were also given the once-over. The standard TD1B modification of cutting 50mm out of the centre section of the exhaust was taken a step further and the total length of the pipe up to the baffle cone was reduced by 115mm. This reduction in tuned length did however cause a problem with the tapers of the diffuser pipe. In order to produce the same volume pipe as had been shown to be necessary on the TD1B, the overall taper angle had to be increased because of the reduction in the length of the pipe. The GP program had shown Yamaha that it was important to maintain as mild a

The rubber-mounted Amal 276 clones were no more effective on the TD1C than they had been on earlier models.

The distinctive bulge of the clutch now mounted on the transmission mainshaft of the TD1C was the decisive factor in giving the machine the reliability it needed to become truly competitve.

taper as possible to prevent rapid expansion and cooling of the exhaust gases in the pipe. Higher temperature gases resulted in higher amplitude scavenging pulses at the exhaust port. A compromise was found in the use of a shallow taper for the first 80 per cent of the diffuser pipe, followed by a doubling of the taper for the last 20 per cent. This was the first step into the world of multi-taper diffuser pipes as used on all modern two-stroke racing engines today.

The 10,000rpm 38bhp TD1C was ready for battle by February 1967, and they were present in great numbers for their annual Daytona walkover. Things nearly went badly wrong as Dick Hammer on a factory X6/T20 Suzuki chased Gary Nixon for most of the race and passed him only to have his rear brake lever drop off forcing him to settle for second place. Yamahas took all the other top six places. During 1967 the X6 was only occasionally able to challenge the TD1C, the best result being at Indianapolis where Gary Nixon again won but was chased home in second and third by Ron Grant and Dick Hammer on X6s. This was the closest Suzuki were to come to challenging the Yamaha hegemony in US 250 class racing that was to last into the 1980s.

The British 1967 racing season was well under way before the first two TD1Cs arrived. Derek Chatterton was one of the first to get one and immediately began stringing together consistent rostrum positions, culminating in his taking the British Championship for that year. Reg Everett

Gary Nixon takes one of the first TD1Cs produced to victory at the Daytona 100 miler in March 1967.

Growing Pains

was still battling away on a TD1B that eventually took him to a well-earned third place in the championship. The meat in the TD1 sandwich was Peter Inchley who rode the socks off his Villiers Starmaker to achieve a number of good results. His second place marked the swansong for competitive two-stroke engines from British manufacturers. Throughout Europe TD1Cs were mixing it with the best machines. Even one or two appeared in Italy, the bastion of lightweight four-stroke machines, but Aermacchi and Benelli managed to hold off the challenge in the national championships.

The year 1968 was to be the year that the TD1C was felt reliable and quick enough to mix it with the big boys in GP racing. As the world watched the feud between Phil Read and Bill Ivy on their V-4 Yamahas, Rodney Gould's performance on a TD1C engine housed in a Bultaco frame was almost overlooked. It was a spectacularly successful season, resulting in three third places, three fourth places and two fifth places. In the British meetings he contended, it was a three-way battle between Derek Chatterton, John Cooper and himself behind Ivy and Read on the V-4s, when they were present. The engine had been brought back with him from California and mounted in a Bultaco frame by Ron Herring. In his capable hands, the TD1C gained a few bhp, so that by the middle of the season it seemed to have the edge on its brothers.

As the 1968 season drew to a close and Yamaha announced their intention of withdrawing from direct factory support for the 250 class, it became clear that the TD1 was poised to take its place as the most competitive Japanese 250 road racer. The FIM was also becoming increasingly involved in 'reducing the cost of racing' by imposing restriction on engine configuration for each class. From 1970, the 250 class would be limited to twin cylinder engines with 6 speeds. The TD1C or its successor would fit the bill nicely. Within ten years of its introduction, the TD1C had evolved from a heavy, gutless, unreliable joke of a racer to a machine that could challenge the best in the world of 250 road racing. The vanquished had become the vanquisher.

4 Heir to the Throne
TD/TR Production Racers 1969–73

When Honda announced their withdrawal from the world of grand prix racing at the start of 1968, they surprised everyone, including Yamaha. Yamaha had expected another titanic struggle between the six-cylinder Honda 250 and the V-4 Yamaha. The lack of works competition in both the 125 and 250 classes, detracted from the prestige accorded to the winner of the world championship. However, the V-4s were little more than a year old and had cost much time, resources and yen in their development. The race budget for 1968 was allocated and the campaign had been planned. It seemed foolish to cancel everything at this late stage, so the factory proceeded with official participation in the 1968 racing season.

TD2 AND TR2

At the same time, Honda's withdrawal had prompted them to consider their future race plans. There was little point in the further development of the V-4s, since they would be illegal from the 1970 season with the FIM regulations that were being drawn up. The TD1C had shown itself to be a capable racing machine at clubman level and with the potential of mixing it with the pukka sub-top GP machines of other factories. By April 1968, a decision had been taken to develop a new generation of production road racing machines, based as always on the road bike series, that would be capable of winning grands prix. The 250 would be joined for the first time by a 350, derived from the second series of street bike 350s, the YR2. The TD2 and the TR2 had been conceived.

The members of the research and development team charged with the design of the machines did not need to spend any time analysing the deficiencies of the TD1C. Most of the one hundred and eighty owners would have characterized the TD1C as having a decent engine in a second-rate chassis. By significantly improving the chassis and doing a little engine development, a much better bike would be produced. In addition, it would be advantageous for as many parts to be interchangeable between the 250 and 350 racers as possible, both to reduce development and production costs and encourage the use of both machines with a largely common set of spares. It was this joint development of 250 and 350 machines by Yamaha that was to lead to GP grids in the 1970s consisting of 75 per cent of the same riders in both classes. By developing the bikes in this way they had made racing cost-effective for both themselves and the riders. The down side was that the racing sometimes appeared to be a re-run of the previous race, with the machines and riders seemingly almost identical. It could be argued that this contributed to the demise of the 350 class in the early 1980s.

Much of the bottom end of the engine was either YDS6-based (crankshaft, crankcase) or unchanged from the TD1C (ignition,

Heir to the Throne

At the 1968 Daytona 100 miler experimentation that led to the TD2 was already evident. Here a TD1C engine in an RD56 chassis.

transmission, basic clutch design). But there were enough changes to the engine to push the power up to over '44bhp' at 11,000rpm. Again the thickness of the piston ring was reduced to 1.0mm to pre-empt any flutter at the higher engine speed.

There were changes to the port dimensions in the cylinder. The TD1C gulley pseudo-port became a fully fledged, crankcase-fed port. This auxiliary transfer port had a flat roof at the cylinder wall junction in contrast to the main transfers that had a 15 degree elevation towards the cylinder head. During the development of the GP machines, Yamaha had stumbled on the power improvement they achieved when flattening the roof of the auxiliary transfer ports. Most likely they believed that by directing the charge streams directly at each other over the piston crown, they were 'washing' away residue exhaust gases that would otherwise be missed by the main stream of charge directed towards the cylinder head. In reality, the power improvement was less due to this than to the fact that the auxiliary port charge was not being lost out of the exhaust in an engine that was in effect being over-scavenged. It was to be another seven years before this was realized and both main and transfer ports given flat roofs. The exhaust port itself was raised by 2mm and gained 1mm in width. The main transfers were widened by 4mm and the inlet port lost 1mm in width as well as the floor of the port rising by 1.5mm These changes only made sense in conjunction with the change of carburation from the flawed and obsolete Amal 276 clones to a fully integrated Mikuni 30mm aluminium round-slide carburettor. The transfer and exhaust port changes increased the effective transfer port area significantly to flow more charge from the crankcases. Slightly smaller inlet ports and shorter inlet timing was possible due to the larger carburettors.

To complement the changes on the inlet and transfer side of the engine, a new exhaust was designed. The main characteristic of the pipe was the move to a long steady taper right from the cylinder junction. The header pipe had a 1.6 degree taper, developing the concept of employing gradual taper in the diffuser section of the pipe to maintain exhaust gas temperature. The overall length of the pipe dropped by 11mm but was still too long as many owners of the TD2 discovered and corrected for themselves. Yamaha also decided to make life easier for the rider by installing their 'tried and tested' Autolube system, first employed on the road bikes at the beginning of the 1960s. What was perceived as a boon for street bike owners was seen as a superfluous gimmick by the riders who almost universally junked the Autolube along with the kickstart mechanism with which all TD2s were provided.

On the chassis side, the street-based frames used on the TD1 series were discarded as totally inadequate. The modest power of the TD1C had vividly demonstrated the shortcomings of the frame; housing the TD2 engine in a similar frame would have been madness. The only other chassis with which Yamaha had race experience was that of the RD56 first employed in 1963. After some initial problems with engine positioning, the RD56 frame had been able to handle the 50 plus bhp of the engine reasonably effectively, so the 44bhp of the TD2 should not be a problem. Thus the classic 'Featherbed' design of the Manx Nortons of the 1950s found its way on to a Japanese production racer of the 1960s. You can't keep a good design down! Suspension front and rear was updated and the double twin leading-shoe brakes from the factory V-4s were used to provide sensational (for the time) front wheel stopping power. Yamaha might not be officially supporting the GP World Championship, but they were not averse to using GP technology to help their customers to racing glory. The finishing touch to this racing package was the re-designed fuel tank and seat which doubled as an oil tank for the Autolube lubrication system. Both fuel tank and fairing were decked out in Yamaha's corporate livery of white with a racing red stripe down the tank and on the face of the fairing.

For the first time, the 250 production racer was joined by a big brother, fast, mean and lean enough to clean up in the 350 class and take the 500 class as well with a bit of luck. The successes of the TD1 had given Yamaha enough confidence to mount a challenge to the larger classes of both clubman and national championships. Although this was the first mass-produced 350 production racer, it was not the first roadster-based 350 that Yamaha had constructed. Back in 1967, Yamaha in the USA had entered a 350 machine based on the newly introduced YR1 350 street bike for the Daytona 200. The AMA had some really strange rules for the machines entered in the 200-miler. The 500 class machines, including 350s, were permitted to use non-standard frames if submitted for approval, but were restricted to a maximum of four gears regardless of the street bike gearbox. Taking advantage of the sanctioned modifications, Yamaha slotted the R1 engine in an RD56 frame, but were required to utilize the standard YR1 forks. Having only received the engine in late 1966, Yamaha had no time to re-design and manufacture the optimum four ratios. Instead they achieved the four speeds by dropping standard first gear, left second and third identical and produced a new top gear with an overall reduction of 4.9:1. The 30mm

The first TD2s started arriving in Europe in April 1969 and they sold like hot cakes. Kent Andersson got one of the first and put it to good use in the GPs.

Mike Duff on the TR1 narrowly ahead of Gary Nixon on the Triumph at the 1968 Daytona 200 miler.

versions of the remote float 'Amals' used on the TD1s were fitted and electronic ignitions were fitted for the first time to Yamaha production racers. Although ready for Daytona, they were far from sorted machines. The gearing was found to be less than satisfactory, first being too high for some of Daytona's slower corners and top too low for the Tri-Oval banking with a tail wind. The top gearing was raised after an engine was over-revved and blew, but the correct gearing was not identified. Six machines were prepared for the race, two of which were ridden by Tony Murphy and Mike Duff. Mike Duff was holding tenth place when the petrol tank split and he was forced to retire. A year later, the TR1s were re-entered in the hands of Yvon du Hamel, Art Baumann, Mike Duff and Phil Read, essentially the same as the 1967 versions, still restricted to four speeds. This was to be Cal Rayborn and Harley-Davidson's year, but du Hamel and Baumann brought the Yamahas home in second and third place, auguring well for the production racer to be available a year later, the TR2.

The TR1 had proven the effectiveness of the RD56-based chassis and had functioned well as a testbed for the TD2 and TR2 machines. The engine of the TR2 was not quite as sophisticated as that of the TD2, in that the auxiliary transfers bore a close resemblance to those used on the TD1C. They were not fed directly from the crankcase, but rather from windows cut in the skirt of the piston. Unfortunately this also weakened the piston skirt and it was quite common for the piston to crack around the window. On the TD1C, the transfers had been open gulleys but the TR2 ports were passages cut into the cylinder wall. For the first time, horizontally split crankcases were adopted for a production racer, for the simple reason that the 350 street bikes also had horizontally split cases. Although the engine appeared very similar to the street bike, there were fewer shared road and race parts than on the 250. Bore and stroke were the same as the YR3, but the engine turned at 9,500rpm, 2,500rpm higher than its street cousin. This was the reason for the different crankshaft, con-rods, single-ring pistons and bearings. The engine was fed with 34mm Mikunis and a multi-diffuser exhaust of similar design to the TD2 was used. Clutch and magneto ignition were borrowed from the TD2, but gear selection was by forks mounted on a rotating drum rather than the rotating pawl system that all the TDs, up to and including the TD2, used. The resulting package was good for a claimed 55bhp and a top speed exceeding 140mph (225km/h).

The first TD2s to be raced competitively were pre-production engines in TD1C frames that were run by Yvon du Hamel in Canada at the end of 1968. The AMA sensibly raised the limit on 500cc class machines to five speeds, making the TR2 totally legal for Daytona. Yamaha looked set to finally claim the most prestigious race in American road-racing. Once again it was to be Cal Rayborn and Harley that thwarted them, the TR2s proving to be fast but fragile as Yvon du Hamel set a qualifying lap record of 150.5mph (242.3km/h), but pulled out of the race after six laps with a partial seizure. Ron Pierce, on another works Yamaha, crashed leaving Mike Duff and Rod Gould to defend Yamaha's honour and they finally finished third and fourth respectively. Ron Grant on the Suzuki 500 XR05 had a good race to take second place behind Rayborn. Traditionally Yamahas took the first three places of the 250 race ridden by du Hamel, Baumann and Gould respectively. Cal Rayborn managed fourth on an A1R Kawasaki racer.

Gould owed his Daytona ride and good result at least partly due to Randy Hall his mechanic for the 1968 season.

> I'd met Randy in Los Angeles during my stay there during the winter of 1967. He'd expressed an interest in coming to Europe and had looked after my TD1C-Bultaco and 500cc Manx Norton. At the end of the 1968 season he had returned to Los Angeles and got a job in the Yamaha race shop, who were pleased to make use of his skill and experience in Europe. He arranged to purchase

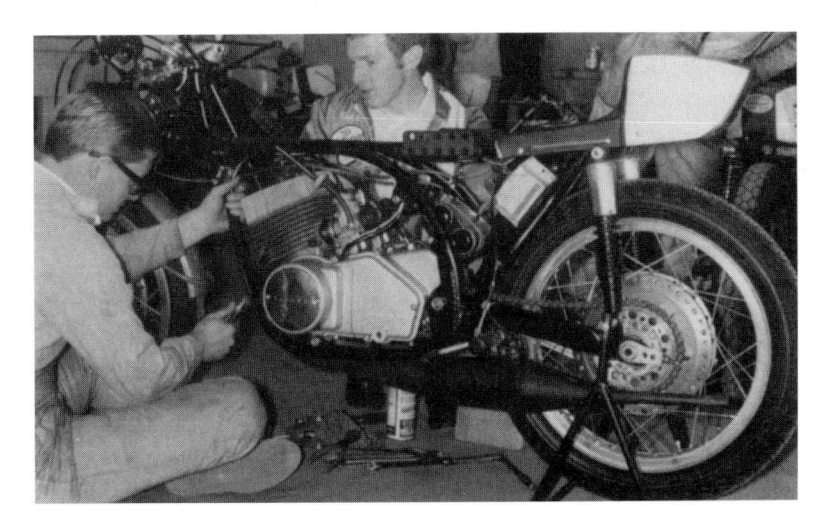

The other secret weapons in the Gould collection was this TR2 and Randy Hall (left) whom he'd also brought back from the USA.

Heir to the Throne

one of the sixteen pairs of TD2 and TR2 machines Yamaha imported for the 1969 Daytona. I should have won Daytona that year. I was running second to Cal Rayborn on the Harley, when the springs holding the exhausts onto the barrel broke. We had them wired on as a precaution to stop the spring popping out, but not to support the pipe if the spring broke. I pulled into the pits, we fitted a new spring, but I could only pull up to fourth place. At the end I was lapping quicker than Rayborn; I'm sure I could have beaten him if only...

After Daytona, we shipped the bikes to Europe where we ran them with spares provided by Yamaha International. We arrived back in Europe in time for the Easter meetings at Brands, Mallory and Oulton and from six starts I won six races with them. Prize money from that weekend payed for the purchase of the two bikes including transportation from America. You can't do that anymore.

I did the GPs on them, but the fuse was a bit short on those early TD2s and especially the TR2s. The heel of the points would break, the cages on the big-end bearings would fail, exhaust springs would break, heads would crack, con-rods would break. We had quite a few holed pistons on the 350. The chassis was good, although the exhausts grounded too easily on the standard

Rod Gould cleaning up at the Easter races of 1969 on the TD2 he'd brought back from Daytona with him. He's shown here racing at Mallory Park.

machines. They were short circuit bikes, anything over 50 racing miles and the chances of failure rose considerably. On the short circuits I was almost unbeatable, winning twenty-one of the twenty-three starts that year. I even managed to beat Ago on the 500MV at the Post-TT race at Mallory Park, the first time he'd been beaten in the UK.

Gould's own experience closely reflected the

Gould was in top form during 1969 at home and abroad. Here he gives Agostini on the MV a run for his money at the Czechoslovakian GP at Brno. Ago was to pass him before the chequered flag.

experiences of most of the first TD2 and TR2 owners. The bikes would reward their owners by an amount directly proportional to the amount of attention they got. A very strict preventative maintenance programme involving crank rebuilds and piston ring replacement after every meeting, and absolutely spot-on setting of the ignition, payed off with good race-long performance. Kent Andersson, an engineer by trade, was the first to reap the benefits of his lavish attention to the engine. On 11 May 1969 at a sun-drenched German GP at Hockenheim, 100,000 spectators saw Kent Andersson win the 250cc race at an average speed of over

The TR2 was not a complex machine and the horizontally split crankcase made the mechanic's life easier than the vertically split TD2.

Kent Andersson becomes the first GP winner on an 'over-the-counter' Yamaha production racer at the 1969 German GP at Hockenheim.

100mph (160km/h) on an over-the-counter TD2 road racer. This ranks as one of the five most important successes in Yamaha's road racing history. They had produced a racing machine freely available to anyone with £900 and the necessary talent that was capable of beating any other machine in its class. It was an astounding achievement.

Suddenly everyone wanted TD2s and TR2s. Supplies in Europe were very limited for the first half of 1969 and many racers imported machines directly from the USA. It soon became clear that good machines that they were, even better performance was possible with a few minor changes. The 250 engine's exhaust side could be improved by widening the exhaust port by 2mm and shortening the diffuser lead-in pipe by 30mm. Raising the exhaust port of the TR2 by 1mm as well as the extra 2mm width helped both the mid-range and top end of the engine, as long as the inlet side of the piston skirt was reduced by 2mm to help flow more charge into the crankcase.

The 1969 season drew to a close with Phil Read achieving a 250 and 350 double in Ago's

Heir to the Throne

After the 1969 250 World Championship on the four-cylinder Benelli, Kel Carruthers switched to Yamaha, starting a relationship that would last for 20 years.

absence at Imola. Kent Andersson finished second in the 250 World Championship to Kel Carruthers, who was riding the four-cylinder four-stroke Benelli, and Giuseppe Vincenzi was third in the 350. There were no updates to either the TD2 or TR2 for 1970, but Yamaha returned to GP racing via Yamaha Motor NV in Holland. Rod Gould and Kent Andersson were contracted to ride the development machines for the TD2B and TR2B that became available in 1971. From the start of the season, the TD2s were fitted with factory six-speed gearboxes, and Kent Andersson had commissioned Stan Hedlund in Sweden to construct an equivalent six-speed transmission for the TR2. From the Yugoslavian GP in May, they used electronic ignition supplied by the Spanish Femsa company and based on the Femsatronic ignition used by Bultaco in 1969. In their contract with Femsa, Yamaha had stipulated that only the factory riders could be provided with the ignition. It was hoped that this strategy would keep Gould and Andersson one step ahead of the opposition. This was only partially successful as, within a couple of races, Kel Carruthers, now also riding a TD2, had commissioned the German Kröber company to produce an equivalent ignition that marked the start of the 'ignition wars' that were to wage for a couple of years until the Femsa became standard on European machines in 1972. The battle for the championship went down the line to the penultimate race of the year at Monza, where Gould's win made him and Yamaha the undisputed champions of the 250cc class.

Yamaha felt everyone should benefit from the development year that Gould and Andersson had spent on the TD2s, and as a result the TD2 and TR2 both underwent minor updates to become the TD2B and TR2B respectively. The cylinder port changes that most TD2 and TR2 owners had

Gould (8), Carruthers (12) and Read (7) fought tooth and nail for victory at Imola in 1970. A win for Carruthers would keep his title hopes alive for the last GP of the year. A win for Gould would mean the title. Gould won.

been applying were made official along with a reworked con-rod small-end to reduce the number of bearing failures owners had been experiencing and a piston with shorter skirt on the inlet face for the TR2 as well as a slightly thinner head gasket to raise the compression ratio. Both machines received exhausts with revised dimensions and frame support brackets and a slight change in the support brackets for the carburettors. Yamaha also took the opportunity to raise the gearing slightly on the 250 for all gears except first, to take advantage of the higher power and torque values the engine was now producing. The rest of the machines were left unchanged with the single exception of the fuel tank. On both models the cap was moved from the right- to the left-hand side of the machine and was changed from a thread release to a spring-loaded pressure release. This more than anything helps to distinguish the 2s from the 2Bs.

Thanks to Yamaha's own racing success with the TD2s and TR2s, the Bs sold like hot cakes. Both 250 and 350 grids were becoming almost exclusively Yamahas at all levels of the sport. Phil Read had spent most of 1970 racing in the UK with occasional GP outings producing sufficiently good results to persuade him to take to the 1971 GP trail, using his own money and with some sponsorship from Castrol. The basis for his challenge were the 2B models, but during the season they evolved into very different machines, thanks to Ferry Brouwer and Helmut Fath. Eric Cheney built the team a new frame from Reynolds 531 tubing saving five pounds in the process, whilst retaining essentially the same geometry as the standard frame. Disc brakes were fitted front and rear, one of the first times they were to be seen on a TD. Rod Quaife was commissioned to produce a six-speed gearbox, squeezing a gear between third and fourth. Kröber ignition was fitted from the start of the season and Girling shock absorbers replaced the standard rear suspension. A steering damper was fitted to quieten down the front end. Within a couple of GPs, Fath had been persuaded to carry out a conversion of the wet clutch, which was quite heavy to operate as standard, to a dry clutch. Brouwer shortened the standard exhaust a little, 10mm coming out of the header pipe and 30mm from the middle section, as well as doing some porting work. The TD2B was transformed into a machine that

Heir to the Throne

The Cheney frame Read used during 1971 to win the 250 title could also take a TR2B 350 engine as shown here.

The first of the three 250 victories that led to Read's world championship in 1971 was here in Hockenheim. This was his first win since the title from Ivy at Monza in 1968.

went even faster than it looked. It was enough for Phil to beat Gould, who had an up and down year, running 1970 models and prototypes for the 1972 production machines. The championship was finally decided at the Spanish GP at Jarama, where Gould's clutch failed. Read came second, winning the World Title, and Jarno Saarinen won his first 250 GP.

In the USA, Kel Carruthers was busy blazing a trail for the TR2B. Making many of the changes that Europeans were also applying to the 350, Carruthers was able to extract around 60bhp from the TR2B engine spinning at over 11,000rpm. He also made some changes to the geometry of the bike by lengthening the swing-arm and moving the engine slightly further forward. The result was a powerful package with handling to match and it enabled Kel to win at Road Atlanta, and take second place at Loudon, Pocono and Ontario. Yamaha were poised for domination of almost all classes of road racing throughout the world.

TD3 AND TR3

In hindsight, it could be said that Yamaha produced one of their best 250s ever in the TD2B. With some minor adjustments to improve reliability it could have been the basis for a generation of machines well into the 1970s. Tuners in particular loved it – it seemed to respond to their efforts exactly as they expected it to. The TD3 successor turned out to be mysteriously difficult to work with, even the factory having trouble with its development. Actually, the TD3 and TR3 were not produced to offer vastly improved performance. This was no longer necessary as the 250 was totally dominant throughout the sport and the 350 was approaching GP dominancy with only the MV left to defeat. The TD3 and TR3 were built as 'cost-reduced' 250 and 350 racers. The trend that had started with the TD2 and TR2 which had shared the same chassis was extended on the TD3 and TR3 by modularizing the engine. By changing the short-stroke dimensions of the 250 from 56 × 50mm to a square 54 × 54mm and the 350 from 61 × 59.6mm to 64 × 54mm, it was possible for both engines to share the same bottom end. Making the most of this change resulted in almost total commonality of the engine, clutch, six-speed transmission and chassis. The clutch was at last dry. The ignition was electronic, European machines were delivered to Amsterdam without the standard Hitachi ignition fitted and Femsa units went in the crates before passage to the importer.

The first of the '3's had an inauspicious baptism of fire at the 1971 Dutch TT when Rod Gould could do no better than fourth place, the TR3 clearly down on top speed, and performance seemed to barely improve throughout the rest of the year. Early in 1972, the rest of the racing world was given the opportunity to evaluate the bike, and its smaller brother the TD3, themselves. On paper they looked like a good deal, particularly with the possibility of using a single machine for both classes. For the 250, the horizontally split crankcases made life easier for the mechanics, which was just as well as the engines were being stripped regularly. Many riders were plagued with detonation problems, the exact cause of which was never really established. The 1.2mm gap that had existed on the old engine at TDC between the cylinder head and the piston crown had been removed and the exhaust pipe had yet more multiple sections in the diffuser. Perhaps the increase in exhaust gas temperature and closer piston crown caused the problem, but for many riders, the TD3 did not match either the reliability or speed of the TD2B. The factory only campaigned the air-cooled '3's in Grand Prix for the second half of 1971, since by 1972 they had the pre-production water-cooled 250s and 350s in use.

At Daytona, Kel Carruthers was clocked at 164mph (264km/h) during practice for the 200-mile (320km) race on a TR3 he had received at the end of 1971. Fastest in practice,

From 1972, the TD3 engine was also split horizontally and became almost indistinguishable from its big brother the TR3.

the bike failed him in the race when the ignition packed up, although Dave Emde on a TR3 led home two other privateers making it a Yamaha 1–2–3. The 250 race was won by Dave Smith running a TD2B, with the official 'works' TD3 of Kenny Roberts, at his first Daytona, coming second. The rest of the US season was more failure than success. Gary Fisher, riding one of Kel's TR3s at Loudon soundly thrashed the Kawasaki H2Rs and Suzuki TR750s, but it and Daytona were Yamaha's only big race wins in 1972. The Harleys were still invincible on the slow circuits where handling and low speed acceleration were so important and the Kawasakis won at Talladega and Atlanta.

In Europe, Jarno Saarinen was the man to beat in 1972. The Yamaha factory team effort that year was a disaster, with six riders all expecting to get the best equipment as they rode for six European importers, but they were getting all the bikes and spares directly from Yamaha Motor NV in Amsterdam. Jarno was able to avoid the political infighting that developed and take the 250 crown after a strong challenge from Renzo Passolini on the new two-stroke Aermacchi. He also gave Agostini and MV such a scare in the 350 class that they hurriedly introduced a 350 four-cylinder bike to replace the triple. It was enough to earn a stay of execution, but 1973 would surely mark the end of MV's hegemony in the 350 class.

There was a six-month delay in the production of the 1973 TZ250, and in order to adhere to the new corporate identification standards of including the capacity in the model code, an extra batch of TD3s were produced with the designation TA250. The A stood for air-cooled.

TA125

There was another TA model introduced during 1973, the TA125. The factory could never

The 1972 TR3 was to follow in the footsteps of its giant-killing predecessor, the TR2.

really make up their mind how to approach the 125 class after their withdrawal of the V-4 in 1968. Initially they had provided a special kit that could be used to convert the YAS1 street bike to race specification. This resulted in a good engine that was not matched by the quality of the chassis. In the UK a number of Yamaha racing dealerships offered their own frames to complement the Genuine Yamaha Tuning race kit. Granby-Yamahas and Padgett-Yamahas largely dominated the 125 class in the UK. Although the YAS3 was first produced in April 1971, it was to be two years before Yamaha released a full racing version of the bike in the TA125. As much of the standard road bike was used as possible, the most notable engine difference being the use of electronic ignition whose rotor was attached to the left-hand end of the crankshaft. As on the street bike, bore and stroke were both 43mm. The chassis was also essentially AS3, with minor modifications to the rear suspension and swing-arm as well as a YDS6 250 road bike front wheel. The resulting machine produced 24bhp at 12,500rpm and was used with some success in clubman racing throughout Europe during 1973 and 1974, although interest in the class in the UK was almost non-existent. Yamaha though were not really interested in this class and they let the TA125 slip away without a water-cooled update.

The lack of impact of the TA125 on the racing world is strongly contrasted by the success of the factory 125 team from its introduction in 1971 through to 1975. It had been the intention that both Rod Gould and Kent Andersson ride the 125 in 1971, but Gould wanted to concentrate on the larger machines. Charles Mortimer was drafted in to ride the bike at the Isle of Man and won his first GP. Riding the bike for the rest of the year, he managed to take fifth place in the championship table. From 1972, Kent Andersson adopted the bike and after coming runner up to Angel Nieto on the Derbi that year, he won two world titles in successive years. The introduction of the Morbidelli in 1975 ended his domination of the class. The 1971 works machine was air-cooled and 43 × 43mm as the TA125 was to be. For 1972, Yamaha produced a short-stroke water-cooled works machine for Mortimer measuring 44 × 41mm and producing approximately 35bhp. The wet clutch was still fitted at the start of the season, but by the end of the season had been replaced by a dry clutch. The front end was lifted from the old V-4 125 racer. From 1973 Kent Andersson was given free reign for his considerable engineering talent and he experimented endlessly with all sorts of port timing and engine configurations including a disc valve behind the engine. It was, and remained, a totally European effort; Yamaha Japan were uninterested in the 125 class.

The development of the TDs from fragile, slow, uncompetitive racing motorcycles into the dominant force in road racing had been a journey for Yamaha in search of omniscience, complete understanding. By the early 1970s they were the acknowledged two-stroke experts of the world. Their 250 and 350 racing machines were the best that money could buy. TDs and TRs had won literally thousands of races in their evolution to excellence for both their riders and Yamaha, every success contributing to the Yamaha legend. The journey was not complete; but the TD had established a foundation that was an unshakeable base to future glory.

Kent Andersson discovered the joys of 125 racing during 1972 and won here in Imatra, Finland.

5 Private Passion

TZ Racers 1973–80

Water is essential to life – all life depends upon it and all living things contain it – and the average person can exist for only three days without it. Water is the coolant that keeps the body at an even temperature and vital parts of the body need it to function effectively. The same is true of a highly tuned racing motorcycle; its heart, the engine, needs to be kept at an even temperature to function effectively. Heat is both the friend and enemy of internal combustion engines, for it is heat that results in the mechanical energy to turn the crankshaft, but it can also be responsible for both seizures and detonation within the cylinder. Only about 25 per cent of the energy produced by combustion is actually converted into mechanical energy. Some is lost through the exhaust, but about 40 per cent of the heat produced by combustion must be absorbed by the cooling system. The more efficient the combustion process, the more power at the crankshaft, but also the more heat that the cooling system must dissipate.

WATER-COOLED RACERS: TZ

Yamaha's own experience in the GP wars of the 1960s, with both air- and water-cooled two-stroke engines, had shown that the limit of effective air-cooling was around 55bhp. Tests had also shown that water-cooling was much easier to fine tune to provide the optimum operating temperature of an engine at 80°C. It was clear that the TR3 was already around 55 bhp and there had been some problems with both detonation and seizures, although these were not limited to the 350. Water-cooling was clearly the next step, and when Gould and Andersson received pre-production TD3s and TR3s, sharp eyes in the paddock noticed the lugs already welded on the frames in anticipation of the radiators that would appear six months later. The first sets of water-cooled cylinders and heads were in the Yamaha workshop at Uithoorn, The Netherlands before the end of 1971. Ferry Brouwer, working for Yamaha by this time, mounted the green-anodized water-pump, cylinders and heads on Mortimer's bike for testing at Zolder at the start of 1972. By the start of the season, the paddock was practically swimming in pre-production water-cooled Yamaha 250 and 350 racers, with Saarinen showing how good they could be by taking the 250 crown and chasing Agostini hard in the 350 class. By September 1972, the production line had been tooled up to produce Yamaha's first water-cooled production racer, the TZ350.

The new machine was very close to the TR3, but with water-cooling. The cylinders were cast in a single block of alloy, with water passages that extended up through the single-piece head. A thermostat was plugged into the cooling system at the front of the head between the two cylinders, intended to keep the temperature at 80°C. A small aluminium radiator was mounted on the down tubes at the front of the frame

Private Passion

Chas Mortimer's pre-production TZ250 in the spring of 1972.

During the second half of the 1972 season, Saarinen was almost invincible in the 250 class. Starting here at the Belgian GP, he took a hat-trick of wins, on his way to the championship title.

underneath the headstock. The water-pump replaced the Autolube pump that the air-cooled models had fitted and was therefore driven by the transmission mainshaft running at 66 per cent engine speed. Water was pumped into the back of the cylinder head, down through the cylinder jacket, past the thermostat, into the radiator and returned to the pump. A rubber O-ring sealed the cylinder head/cylinder face. A temperature gauge was plugged into the thermostat housing and the water temperature displayed on a meter next to the rev counter. The owner's manual warned ominously that 'The engine on the TZ350 is capable of producing the highest output when the water in the radiator is around 80°C. It is necessary to be careful when the water temperature

Private Passion

meter reads above 85°C.'

Apart from the cooling, there were very few changes to the machine. There was no real modification to the port timing in the new cylinders, but there was a slight change to the exit angle from the transfer ports. The auxiliaries were lowered slightly from the 45 degrees found on the TR3 to 30 degrees. Yamaha had still not discovered the most effective way of scavenging the exhaust gases from a piston-port cylinder. Complementing this change was a slightly greater volume exhaust pipe whose tuned length was also slightly longer, lowering the peak power by a few hundred revs. This change would have increased the scavenging efficiency of the exhaust pipe and less fresh charge would have been lost out the exhaust from the auxiliary transfers. A small step in the right direction. The TZ350 put out about 65bhp at the crankshaft at 10,500rpm, and most important of all it kept on giving close to 65bhp throughout the race. The short fuses had definitely been lengthened.

The Yamaha race shop in the corporate research and development section in Hammamatsu was a busy place in the second halves of 1972 and 1973. There were a lot of projects on the table. The water-cooled production racers were being designed alongside Jarno Saarinen's OW19 500cc MV beater and the prototype TZ700s were undergoing development. The TZ350 was given the highest priority as Yamaha sensed another Daytona 200 victory, this time hopefully from the official factory team led by Kel Carruthers. Yamaha pulled out all the stops for this one, even contracting Saarinen to come over for the 200 to join Kèl and Kenny Roberts in the official team. Vince French, newly appointed to prepare Jarno's 250 and 350 machines in 1973 after a year working for Rod Gould, remembers the tension before the race:

> Both Jarno and myself were quite apprehensive as we left Amsterdam for Daytona. It was his first race in the USA, on bikes prepared by someone else, with a mechanic he'd not worked with before. I was very con-

Ago on the MV still leading Saarinen at the 1972 West German GP, but it was Saarinen who took the flag. This was the first time Ago had finished behind another rider in a 350 GP since the East German GP of 1967.

64

scious of working for Jarno, a very good engineer in his own right, and I was worried he would be critical of my work. He'd never had a mechanic work for him before, he'd always prepared the machines himself, so it was inevitable that he was going to get his hands dirty, was going to tinker. During preparation for Daytona, I was trying to work out how the vent pipes for the fuel tank should be located to enable the one-way valve to work. I was playing around with it, not getting it right, and Jarno was watching me out of the corner of his eye. He came over, gave me a wink and said 'If we do it this way, I think you'll find it will sit right', which of course it did. He probably had it sussed out ten minutes before but he'd had the grace to let me exhaust my own ideas first. It really put me at ease, knowing he could help when needed but was prepared to let me have a go first.

Jarno was very, very critical on cranks, so we'd brought a few with us in case the US material did not meet his standards. Some material had also been sent from Japan, and we went through quite a few cranks before he found one that was acceptable, checking for machining of the flywheels, end float, putting them on the dials. In the end, we used one of Kel's, but Jarno was someone who needed to be involved in making that sort of decision. Of course we won at Daytona and that established our good working relationship.

Back in Europe Saarinen won the Imola 200 miler on another TZ350, before dropping the 350 to concentrate on the 250 and 500cc classes. Tepi Lansivouri, a close Finnish friend of Saarinen's, concentrated on attacking the MVs of Agostini and Phil Read, beating them both at Brno and Anderstop. But the Yamaha was not quite reliable enough to finally dethrone MV from the 350 class, though it did force them into drastic action. For the first time since the titanic battles with Honda in the 1960s, there was the forgotten sight of MV engine blocks being opened in the grass fields accurately called 'Paddocks' at the GP circuits. For the previous five years, the MVs had been taken out of the truck, dusted off, filled with oil and fuel, raced, garlanded and loaded back into the truck. MV's days were numbered.

By June 1973, the water-cooled 250 was also rolling off the production line, with

By the summer of 1973, the water-cooled TZ250 was available for all.

exactly the same changes that its bigger brother had undergone. The engine was left virtually unchanged except for a slight widening of the main transfer ports. Crankshaft main bearings and small-end bearings were upgraded, but the rest of the changes were associated with the move to water-cooling. The slight improvement in scavenging efficiency resulted in a claimed 54bhp (an increase of 2bhp) at 11,000rpm (an increase in 200rpm).

The change to water-cooling brought an extra problem for the GP team mechanics, as experienced for the first time by the Yamaha team in 1972. Until the riders knocked some sense into the FIM during the 1980s, it was a tradition to start the racing season early and in countries likely to suffer quite bad weather during the spring; 1972 was no exception as the GP circus got rolling at the Nürburgring on 30 April. Vince French recalls his first GP as Gould's mechanic:

> We'd been to Mettet to give the new bikes a shake-down. The weather was horrible; cold and wet. Then it was on to Nürburgring for the GP. Typically, the bikes were allocated the grass field outside the main paddock as the organizers had allocated the garages to the supporting German sports car competitors. It was real brass monkeys weather. It snowed during practice. It froze during the night before race day, and we had been wise enough, or was it lucky enough, that our bike's coolant had been drained for a rebuild. Some of the other Yamaha team members were not so lucky, and the castings were pretty thin as they found out next morning from the cracks in the heads. The old Araldite came in useful that day ...

Saarinen would undoubtedly have won the 250 title again in 1973, as the first three races were all straight wins, with team-mate Hideo Kanaya following him home each time. Then came Monza and the infamous crash during the 250 race that resulted in the deaths of Renzo Passolini and Saarinen himself. Tepi Lansivouri was thrown in at the deep end, barely recovered from the shock and grief at the loss of his close friend. Yamaha thought that new machines would help and they gave Tepi the OW16 and OW17, 250 and 350 racers respectively. That

The ultra-lightweight 350 magnesium special, the OW17, just before its debut ride by Lansivouri at the 1973 Dutch TT.

Private Passion

The OW17 looked like winning first time out in Holland, until a gearbox full of neutrals dropped Lansivouri back to third place.

Dieter Braun's second place at Imatra clinched the 1973 250 title for him in a season started on a TD3 engine and completed with a TZ250 unit.

Private Passion

the OW16 was a magnesium special was clear to everyone, but in fact the internal design was also totally different, with the engine running backwards and an extra jackshaft driven from the centre of the crankshaft. It was light and powerful, almost too much so. Ferry Brouwer, looking after the 250 for Tepi, recalls strapping a spare crankshaft inside the front fairing after Lansivouri almost flipped the bike during practice in Finland. The crankcases were very fragile, splitting frequently and it wasn't the success Yamaha had hoped for. Dieter Braun went on to take the title in 1973, riding a TD3 at the start of the season and switching to the TZ from the Czechoslovakian GP at Brno. Braun was helped out by one Sepp Schlögel, the man behind the success of many German riders over the last 20 years.

DEVELOPMENT OF THE TZ250 AND TZ350

A and B models

Both the 250 and the 350 were to mark time for three years as the 1974 A models, and the 1975 B models. The 350 was by now the mainstay of not only 350 class racing in Europe but also the 500 class where 351, 352, 354, etc. Yamahas fought out re-runs of the 350 race. This was the heyday of the 350 racer in Europe, the direct opposite of what was happening in the USA, where it was the TZ750 that was selling like hot cakes. The 350 had nowhere to go in US road racing and became a racing tool for non-American markets. MV were finally ousted from their domination of the 350 class by, ironically enough, Giacomo Agostini, who had switched camps to get to ride the OW19 500cc GP racer. Ago's 1974 350 was the OW17 Tepi Lansivouri had been provided with from the 1973 Dutch TT. It was a light-

A year after its debut, the Lansivouri 1974 works 350 had changed little apart from the double front disc brakes.

This is the pretty standard looking TZ250 engine that Kenny Roberts rode into third place at his first GP in Holland in 1974.

It was natural talent that took Roberts to the third place at Assen in 1974, but the double disc brake up front must have helped some.

68

weight version of the standard TZ350, with magnesium crankcases and carburettors, lighter flywheels, transmission shafts and frame. It suffered from chronic vibration resulting in persistent cracking of the frame and, at the start of the season, crank failure. Based in Bologna for the early season Italian Nationals, Ago's team worked closely with local craftsmen to produce a less self-destructive crankshaft. Agostini's 350 World Championship drive benefited enormously from this early season work. The bike's 29lb (13kg) weight advantage over a standard TZ made it handle and accelerate a lot better, but the standard TZ350 had no real top speed disadvantage. MV had abandoned the 350 class in a last ditch attempt to fend off the Agostini/Yamaha challenge in the 500 class, and Ago was able to cruise to a comfortable 14th world title. Yamaha abandoned direct support for the 250 class, allowing Walter Villa to keep ahead of the horde of private TZs to win the first of three world titles on the two-stroke Harley-Davidson twin that they had inherited when the company bought Aermacchi.

C models

The year 1975 was to be a mixed one for Agostini, his life being complicated considerably by the arrival in Europe of 19-year-old Venezuelan Johnny Cecotto, quickly dubbed the 'Caracas Kid' by the British press. With standard scruffy TZ250s and 350s and a riding style that always seemed to be on the verge of disaster, Cecotto set the 250 and 350 classes alight. Agostini, in the twilight of his racing career and concentrating on the 500 class was not able to match the aggressive style of the Venezuelan in the 350 class, despite riding a pre-production TZ350C monoshock machine. By mid-season, Cecotto's performances and his sponsor's connections combined to persuade Yamaha to provide him with the monoshock 350 that factory rider Kanaya had been using to perform final debugging of the machine before full production of the TZ350C started. Cecotto also received a monoshock chassis for the 250 class but it was too late to prevent Villa's second 250 title. By the end of the season, the 250 and 350 TZs being ridden by Yamaha's (semi-) factory riders were housed in monoshock chassis. In November 1975, Yamaha called a press conference with both good and bad news; the bad news was their withdrawal from GP racing for 1976, the good news was the announcement of the new TZ250C and TZ350C with a monoshock chassis.

The monoshock system had originated in the Yamaha Motocross team nearly three years earlier. Although not an especially novel concept, having been applied in very similar form to Vincent street motorcycles in the 1950s, Yamaha adopted and developed the design to make it their own. It became a trademark, the perceived advantage of the design perhaps exceeding that obtained in reality. Yamaha had bought the patent for the design from Belgian Lucien Tilkens, as they embarked on their crusade to dominate motocross, as they had in road racing. By 1973, works 125 and 250 machines were fitted with the system and Hakan Andersson won Yamaha's first motocross world championship riding a monoshock YZ250. Despite mixed results in 1974 and 1975, Yamaha decided that all their production racers, motocross and road racing, would be fitted with the monoshock system for the 1976 model year.

The monoshock system offered two different advantages; increased chassis stiffness and more rear wheel travel. The increased stiffness arose from the better triangulation of the rear suspension. The conventional oval-section rear swing arm was enhanced by two tubes running at approximately 45 degrees up from the rear wheel axis to a

Private Passion

The monoshock chassis became available to all with the release of the TZ350C early in 1976.

point directly above the swing-arm pivot. Bushes were welded to these tubes at the point that they were joined by two tubes running up from the swing arm-pivot and the joint was strengthened by extra gusseting. These bushes supported the lower end of the extra long De Carbon suspension unit that ran up over the engine to be bolted to flanges welded to the headstock. Extra bracing tubes ran between the two upper tubes on the swing arm just behind the bushes for the lower mounting point of the shock. Finally, short tubes ran between the upper tubes and the swing arm itself. The triangulated design offered better torsional and lateral stiffness when compared to a conventional twin-shock design. This was of greatest benefit to the road-racing machines. The increased rear wheel travel was of less interest to road racers than to their dirtbike riding cousins who were otherwise forced to use bikes with ridiculously high saddle heights.

Of interest to both types of riders was the increased adjustability of the new rear shock. Spring pre-load had been available to riders on their Girlings in the past, but now it was also possible to 'dial in' the size of the orifice within the shock providing bump damping. The rebound damping could not be adjusted. A total of twenty different positions could be selected, and although very unsophisticated compared with the shocks available in the 1990s, it was a welcome addition to the 'tunability' of the production road racer.

The rear suspension was not the only chassis change that the Cs underwent. At last the double twin leading shoe drum brake that had so dominated the front end of the production racers since 1969 was retired in favour of a disc brake, already common on the race track and becoming increasingly common on street machines. Although light alloy discs were used for the brakes, Yamaha did not go the full hog in reducing unsprung weight and stuck to the tried and tested original TZ750 iron caliper. New front forks were fitted to match the increased rear-end travel available and to improve the damping of the front end.

Although not so clearly obvious, some

Private Passion

The TZ250C represented the best bargain racer Yamaha ever offered for sale. Only the curve of the exhaust pipes enable it to be distinguished from its bigger brother, the TZ350C.

changes were made to the 250 engine that seemed small at the time, but were to unlock the door to a significant increase in power from the 250 engine within a few years. Previous 250 engines had flat-roofed primary transfer ports and secondary transfer ports directed upwards at an angle between 5 and 10 degrees. The thinking behind this was that the fresh charge blew the spent gas out of the cylinder, leaving the charge coming from the primary transfers to fill the cylinder with pure undiluted petroil vapour. A nice theory, that turned out to be inaccurate in one vital area. The spent gases could be very effectively removed from the cylinder by the effects of the exhaust pipe alone, enabling the charge from the secondary transfers to join that from the primary transfers in a big fat power-rich column of rising gas. With the TZ250C, both primary and secondary ports were flat-roofed. As the new scavenging system was directing fresh charge to the centre of the cylinder rather than towards the exhaust port, Yamaha thought it would need slightly longer to perform its more efficient disposal of the spent gases. Consequently they raised the height of the exhaust port by 2mm, considerably reducing the exhaust open duration period. To complement this they reduced the tuned length of the exhaust pipe by 40mm. Yamaha were going for mid-range power at the expense of the top end. The 350's exhaust design was distinctly strange with the pipe coming straight and true out of the cylinder castings with a sharp elbow at the end of the first diffuser section of the pipe and running straight and true back under the engine. It was already becoming a problem to fit the pipe on the bike. With the new longer tuned length of the pipe, the preferred smoothly bent pipe would have extended almost to the extreme back of the rear wheel, and that was without a muffler. Rumblings within the corridors of the FIM, suggested an imminent crackdown on noise emissions which would have necessitated the use of 'stinger' mufflers. FIM regulations also forbade the exhaust extending beyond the rear of the bike, hence an exhaust design that slightly compromised its efficiency.

The rest of the machines were unaltered except for a few detail changes to components that had demonstrated unreliability. As if to make up for the disappointment many had felt by their retirement from the GPs, Yamaha priced the C series machines at incredibly low figures of £1,500 for the 250 and £1,550 for its big brother. As if that wasn't ridiculous enough, both machines came with £500 worth of spares. Truly, racing had never been as cheap as in the mid-70s. As always the Cs were the GP riders' tools for the 250 and 350 classes, but without factory support they were unable to defeat Walter Villa on the Harleys. Johnny Cecotto made a promising start to the season, but after an unsuccessful attempt to get the Yugoslavian GP at Opatija boycotted on safety grounds, his season fell apart. In the 250 class only Takazumi Katayama, a young Japanese racer riding if anything more aggressively than Cecotto had in 1975, came close to the Harleys of Bonera and Villa, without managing to beat either of them at any of the GPs. So two world championship runners-up were the best Yamaha could do in 1976.

D and E models

The 250 and 350 were to remain largely unchanged through the D and E models of 1977 and 1978, as the attention of Yamaha's racing department was concentrated on their re-entry into the 500 class. There were official updates to the engine included in the D model that reflected the modifications owners had been making to the Cs during 1976. The philosophy of improving mid-range power at the expense of the top end, benefited inexperienced and less-skilled riders, but the top national and international riders wanted power at almost any cost. The narrowing of the power band by a few hundred revs would not have worried them. Consequently, exhaust ports were raised by 2mm on the TZ250D and some riders took another 1mm off the roof, but the exhaust pipes were not updated. No other significant changes were made. Yamaha's lack of attention to the TZs was beginning to show in the 250 class, where Katayama could do no better than a single GP win in Spain and fourth place in the World Championship that year. Three Italian rider/machine pairs were too strong for him: Mario Lega on the Morbidelli designed by Jörg Muller, Walter Villa on the Harley and team-mate Franco Uncini. Fortunately the 350 turned out to be more successful, but that was more due to Yamaha Motor NV in Amsterdam than Yamaha HQ in Japan. Thanks to them Katayama was able to use both a conventional 350D twin and a three-cylinder 350 they had built during the winter of 1976–77.

The 350-3 came into existence via the eccentric genius of 1970s sidecar racer Rudi Kurth. The arrival of the TZ750 engine had prompted many sidecar drivers to develop their own 'TZ500' four-cylinder engines using the crankcases from the 750 and 250 cylinders. Kurth's compatriot Rolf Biland was one of these, but typically Kurth decided to base his 500 on the TZ350. Essentially, he added half a 350 engine to the left-hand side of the engine and reduced the stroke from the standard 64mm of the 350 to 51.8mm. Housed in a monocoque chassis with hub-centre steering, the machine produced some good results in 1976. Secrecy was a word that did not figure in Rudi Kurth's dictionary and he would tell anyone wanting to know, exactly what he had done and how. In Holland, Ferry Brouwer and Jerry van der Heiden put together a solo 500 racer that was used by Dutch national rider Kees van der Kruys. In Amsterdam, Minoru Tanaka, head of the racing department at Yamaha Motor NV, saw the potential to develop a 350 three-cylinder in the same way, but based on a 250cc block. Two engineering experts were available to assist, Kent Andersson, no longer a rider but still contracted as a consultant by Yamaha, and a young South African toolmaker named Trevor Tilbury. Yamoto, the Italian Yamaha importer, were prepared to finance the work if it would help Agostini to yet another world title, although Tanaka's main goal was to help Takazumi Katayama to the first world title for a

Private Passion

Japanese rider.

It turned out to be a three-man project, Kent Andersson experimented with port timing on a single cylinder he had rigged up at Amsterdam and back home in Sweden, while Trevor Tilbury was the primary constructor of the engine with assistance from Rudi Kurth who developed a welding technique that prevented damage to the cylinder bores when the extra cylinder was welded on. Hoeckle were commissioned to produce a 120 degree crank and Kröber a 120 degree ignition. Katayama's first race bike was slotted into a standard TZ350D frame with the extra cylinder protruding on the left-hand side of the engine. It turned out that the ignition was very vulnerable on fast left-handers, grinding along the tarmac. During the season both Spondon and Nico Bakker frames were used to centre the engine. Nico Bakker was commissioned to frame Agostini's bike from the start.

The bike's baptism of fire was at Mettet in April 1977, where Katayama came home third in a combined 350 and 500 race despite stopping to change plugs after spluttering round the first lap. The GP debut for the machine was at Hockenheim on 8 May, and it turned out to be a glorious 1–2 for Katayama and Agostini. This was one of Ago's best races from his twilight years, as, almost last off the grid because the Yamaha refused to pick up cleanly, he carved through the field to finish 15 seconds behind Katayama, knocking 2.5 seconds off the lap record in the process. This was to be Ago's only decent 350 result all year, whereas Katayama went on to win the 350 title, using the 350-3 to take the title in the Finnish GP at Imatra. Unfortunately, Japan was livid about this private European venture. They saw it as both a waste of money and a distraction from the link they were about to start building between the TZ350 and the street RD350LC that was to go on sale at the start of 1980. There were plenty of plans for the 1978 season but no money was available and so the bike inevitably slipped into history.

Those disappointed with the 1977 D series production bikes would not have been too pleased with the E series of 1978. These were just a new batch of the D models, with no changes to either chassis or engines.

Agostini had only a single good race on his three-cylinder 350, when he tore through the field to finish second to Katayama at the West German GP of 1977.

Private Passion

The ultimate development version of the three-cylinder 350 was taken to the 1978 Argentinian GP by Katayama but not raced. It was not to be seen again.

down far more acutely to the back of the engine, now possible because of the extra width to prevent the tubes fouling any of the engine components. The single rear shock absorber could now squeeze over, rather than under, the bracing tubes passing orthogonally across the engine. The result was a smaller, compacter motorcycle, some 35 to 45lb (15 to 20kg) lighter, with reduced aerodynamic resistance and very good stability. It was used almost universally by all the top US riders including Steve Baker, Kenny Roberts, Gary Nixon and Ron Pierce in 1977 and upcoming Randy Mamola in 1978. In Europe, Yamaha's most favoured sons, Finn Pekka Nurmi and Katayama, were given similar frames for the GPs.

F models

When the F models appeared in 1979, no-one was surprised that they featured the Lowboy chassis. Accompanying the new frame was a new aluminium rear swing arm, with beautiful buttery-smooth welds and a true box section. The pipes that had been used to brace the upper and lower arms were replaced by a massive aluminium plate which doubled as protection for the rear of the engine from dirt thrown up by the rear wheel. In a clear move to save weight on the new machines, aluminium calipers were used for the disc brakes, the rear disc also shrinking in diameter by 30mm and undergoing a minor re-design to simplify its attachment to the swing arm. The front end also underwent minor re-work, with new fork sliders and revised internal pistons as well as a drop in caster from 27.5 degrees to 26 degrees attempting to cure the front end patter for which the TZs had earned a reputation during the second half of the 1970s. The changes in the chassis led to a drop of 20lb (10kg) in weight for both the 250 and 350.

Yamaha was still far too busy winning back the 500cc crown and establishing their four-stroke line of street machines to update the production racers. Inevitably, in the absence of updates from the factory, top riders and tuners took the matter into their own hands. From 1977, more and more riders swapped the standard chassis for Bimota, Bakker, Spondon, Maxton and other frames. Many of them were very similar in layout to the standard TZ frame, with subtle differences to the engine position, a longer swing arm or conical steering head bearings. Most were offered a weight saving through the use of lighter alloy pipe, but not all of them were that much better.

One of the frames that was a clear improvement was the 'Lowboy' frame that originated in California at the start of 1977. It bore a very strong resemblance to the TZ750D or 'OW31' frame that had heralded the arrival of the TZ750 in the monoshock club. The frame design did not deviate so much from that of the standard frame. The secret lay in splaying the upper pipes of the frame behind the headstock far enough apart to allow the petrol tank to slip down between them. In addition these tubes ran

Kenny Roberts had the talent but not the resources to take the 250 as well as the 500 crown in 1978. He took two wins and two seconds in the four races he finished.

For the 250, the chassis represented the bulk of the changes that were made. Within the engine, there were some minor modifications on the carburettor and a new con-rod was fitted to the same crankshaft, but it *did* get a new exhaust of significantly different proportions. Most noticeably, the centre section of the pipe shrank by 35mm, the diameter of the pipe grew significantly, and the diffuser part now consisted of four different sections. The increase in power that this new pipe provided indicated that Yamaha had at last found the correct design to harness the resonances at play within the pipe. This was the first of a new generation of pipes that were of far greater volume than those used in the past and which were now beginning to play a full role in scavenging the engine of spent exhaust gases.

The 350 on the other hand underwent a major engine update. The four-port engine was turned into a six-port engine by the simple expedient of placing a bridge in the centre of the main transfer port. The transfer port closest to the exhaust port opened first, followed 0.1mm later by the middle port and, after another 0.1mm, the last port. The roofs of the ports were not as flat as those found on the 250, the two outer ports having an elevation of 8 degrees, the middle port just 5 degrees. The slower speed of the 350 engine permitted the scavenging stream of fresh charge to be directed slightly higher over the piston crown without it being lost out of the exhaust. The inlet side of the piston also lost 3mm, opening the port earlier and closing it later. The exhaust pipe on the 350 had now grown to unmanageable proportions forcing the left and right pipes to pass over each other underneath the engine in order to provide the correct tuned length. Even so the right-hand muffler only just kept within the length of the bike as measured from wheel to wheel. The Mikuni carburettors grew to 38mm and were fitted with a power jet that sat in the inlet tract. This jet only provided extra fuel when the engine speed exceeded 7,500rpm, boosting top-end performance. The factory claimed a power output of 72bhp at 11,000rpm, an increase of 8bhp over that of the 'E' model.

Although both the 'F' models were

The total re-design Yamaha performed on the last of the TZ350s kept the 1979 model competive through to the very last 350 class GP three years later.

improvements on their predecessors and gratefully received by many riders, they were not able to de-throne the Kawasaki tandem twins that had taken over the domination of these classes in the GPs from Harley-Davidson. In 1978 and 1979, Kork Ballington did the double and it was only in 1980 that a Bimota-framed Yamaha-engined 350, Jon Ekerold's Solitude, managed to defeat the Kawasaki, now in the hands of German Anton Mang. The chassis update that the 'F' models had undergone was not enough for the top GP riders who found the Bimota and Bakker frames less prone to the front wheel chatter still present on the standard machine.

The 350 was to be produced for another two years in the forms of the 350G and 350H, but they did not differ substantially from the 1979 'F' model. By 1982, the decision of the FIM to drop the 350 class from the GP calendar was confirmed and on 26 September 1982, the last 350 World Championship race was ridden. As if mourning the loss of this class, the heavens opened just before the start of the race, turning it into a miserable seventeen laps of Hockenheim. Favourite to win was Anton Mang on the Kawasaki, but surprisingly it was novice Manfred Herweh on a standard TZ350, riding in only his third GP who won this last 350 race. After all that the 350 had done for the sport of motorcycle racing, it was right and proper that the pocket superbike, ubiquitous mainstay of the class, should be winning even at the end.

TZ250G

The 250, fortunately, still had a healthy future ahead of it. Yamaha felt that the 'F' could do with some improvement in two areas, better handling and more power. The 'F' was still a chatter-prone machine despite the new Lowboy frame and Yamaha thought they might be able to go some way to clearing it up with a stiffer front fork. Both slider and outer tube grew a single millimetre in diameter. That was the extent of the chassis changes made to the 'G'. Not many changes were made to the engine either, but they had a much more significant effect on the character of the bike than the fork change. Having proved itself effective on the 350F, the Powerjet system was applied to the 250, with exactly the same goal in mind, top-end power. Piston and cylinder were also updated to extract the maximum power possible without a major engine re-design. Most drastic change of all came on the inlet side of the cylinder. The inlet timing period was drastically increased by the removal of 8mm from the piston skirt next to the inlet port. Realizing the effect this would have on the longevity of the piston, a supporting bridglet poked down 3.8mm from the centre of the port roof to give the piston some support. Another 4mm was added to the width of the port. All these changes had but one goal, stuffing as much charge in to the crankcase as possible. The total width of the transfers was increased by 1mm as was the width of the exhaust port. The wall of the TZ250G cylinder looked like a Swiss cheese.

Riders developed a love/hate relationship with the TZ250G. They loved the power of the engine; it was again 3–4bhp more powerful than the 'F' model. They were ambivalent about the handling, learning to live with the aggressive style that worked most effectively with the chassis. They hated the expense caused by frequent piston and crankshaft failure, a legacy of the radical inlet timing and power production that had tipped the engine over the razor edge into unreliability. The only effective cure was a complete rebuild after every race, although Hoeckle crankshafts, considered expensive necessities by TZ GP riders for some years, did help.

This engine represented the end of the development road for a 54 × 54mm dimensioned, piston-ported twin with single-cylinder casting. There was no more wall for enlarging the ports, the exhaust was allowing the engine to rev over 12,000rpm so was clearly effective, the castings did not permit further experimentation with the transfer passages, and contemporary reed valve technology was immature when applied to a 250. This engine had nowhere to go. Surprisingly, it ended up being used far longer than the single year Yamaha had intended and it remained competitive in the 250 class for a couple of years, despite the updates brought by later models. The cylinders were adopted by many international and national sidecar crews as it allowed them to easily campaign both 750 and 500 classes with the minimum of work on the engine.

As Yamaha stepped into the 1980s, they still dominated 250cc class racing at national level throughout the world. The TZ250G was the only 'over-the-counter' production racer available, enabling national titles to be won and good, though not winning, finishes in the GPs. During the second half of the 1970s, development of the 500cc GP machines had first call on Yamaha's limited race budget, but rather than abandon the 250 and 350 machines altogether, steady evolution had produced excellent machines and many successful and grateful owners. The challenge Yamaha faced was to maintain this customer base and their own reputation for winning machinery in the maelstrom of competition that was to characterize the 1980s.

6 The Crown Jewels
Piston-Ported 500s 1973–81

When Giacomo Agostini denied Mike Hailwood the 1967 500cc World Championship, a lot of Japanese were happy and most of them worked for Yamaha. MV had held off Honda's no-holds-barred challenge for the last class yet to be won by a Japanese manufacturer. When Honda announced that they were to cease their costly struggle for GP supremacy, thus shelving any further work on the four-cylinder 50, six-cylinder 125, eight-cylinder 250 and 350 and six-cylinder 500, these same people would have sighed with relief. Honda had retired with a burning ambition unfulfilled: the 500cc rider's crown.

Honda's challenge to MV had experienced early fruits when in 1962 Jim Redman had snatched away the 350 title from the Italians in whose hands it had been gathering dust for four years. He had used the 285cc bored-out 250, encoded the RC170, to surprise Mike Hailwood on an obsolete and heavy machine that had undergone no development work since the end of the golden era of Italian racing in the 1950s. Later the RC170 was enlarged to 335cc and

Jarno Saarinen stands guard over his new plaything, the OW19, prior to his winning debut ride at the 1973 French GP.

eventually 349cc as MV fought back with re-designed three-cylinder, four-valve machines through the early 1960s. In the end it was only Honda's retirement that brought the 350 title back to Italy. If the castings of the 1962 RC163 250 had permitted it, this machine would probably have been able to take the 500 titles in the years before MV realized the danger of an imminent Honda onslaught in the 500 class. Their loss of the 350 title was enough to warn them and enable them to produce a machine that ironically had many of the characteristics of the RC170 that had enabled it to take the title in the first place. It was compact, light, moderately powerful and a sheer joy to ride in comparison with the Honda four that Mike Hailwood bravely manhandled to within an ace of the 1967 title. Perhaps it was a case of the dyno blindness that also seemed to affect Honda's Formula One cars at the time, but the RC181 500 was an engine on wheels. In contrast to the well-balanced 250 and 350 six-cylinder machines, the 500 was let down by its totally inadequate chassis. Hailwood's own attempts to improve the chassis were met by sullen reprimands from Japan who had yet to learn that the 'loss of face' experienced by bringing in external expertise was not as painful as that experienced by failure.

Honda left the GP world in 1968 in order to start to build the basis for their worldwide automobile industry. MV went back to easy 'race and win' GPs in both the 350 and 500 classes, and the holy grail of the 500cc World Title remained in European hands. In 1968, Yamaha was the only Japanese company with a pukka factory team in Europe and it would have been possible for them to have mounted a successful 350 challenge to MV based on the RD05, but the already marginal chassis could not have handled 350cc power levels without a major re-design. Anyway, Yamaha, as Honda, had other things on their mind and withdrawal at the end of 1968 was the result. Yamaha decided to bide their time.

The next three years were years of racing on the cheap for Yamaha as every yen and man-hour went into establishing the four-stroke flagship, the XS-1 clone of the Triumph Bonneville and its cousins. Busy improving the breed of two-stroke twins by racing them, they were happy to see the progress being made, especially in the 350 class, where the TRs were clearly closing the gap with MV. This, the Italians could of course also see for themselves and they were not going to let the upset of the early 1960s repeat itself in the early 1970s. By the end of 1971 MV had already produced a new four-cylinder version of the 350 in a first attempt to hold off the challenge that was beginning to come from the TR. It was to be an up-and-down year in 1972, but by winning six of the twelve GPs, Agostini was able to prolong his 350 title for another year. They stayed one step ahead of Yamaha this time. During 1972, a four-cylinder 500 became available, in anticipation of the arrival of a new contender in the prestigious 500 class. With the road four-strokes established and a racing feedback loop with the TDs and TRs in place, Yamaha decided it was time to hit the headlines again with a high-profile attack on the Blue Riband class of motorcycle road-racing. On 22 February 1973, Yamaha announced that Jarno Saarinen would be contesting the 500cc GPs on a 500cc four-cylinder two-stroke. The die had been cast.

OW19

The designs of both the YZR500, or OW19 as it was officially designated by the factory, and the TZ700 were put to paper at the end of 1971. Head of the small team responsible for their design was Naito, with chief designer Takasi Matsui. The designs of the machines were very similar, both water-cooled,

The Crown Jewels

The OW19 engine in all its sandcast glory. The reed valves were a comparatively late addition to the engine when Saarinen's test revealed it to be too peaky without.

disc-braked, with horizontally split crankcases, two separate crankshafts, and crank pins spaced at 180 degrees, driving the transmission via a jackshaft behind the cranks. Both machines were Yamaha's first road racers to be fitted with reed valves. Yamaha adopted this induction system in an attempt at fixing the perennial problem of piston ported two-strokes, namely the difficulty of generating both low-end and high-end power. High-end power was dependent on a well-filled crankcase which in turn required an inlet port of such dimensions that at low engine speeds, with ineffective resonances in the inlet tract, much of the charge would be blown back out through the inlet tract. This resulted in loss of charge as well as disturbance of the smooth air flow through the carburettor. It was especially important for the lugging power needed on motocross machines, but it was also felt to be a useful weapon in the battle against the four-stroke MV with its wide power band and good acceleration away from slow corners. The price to pay for this improved low-end power would be a loss of some power at the top end due to the restriction in the inlet tract caused by the reed valve, even when fully open. The 'doubled-up' 250, with the potential for well over 100bhp, could afford the loss of a few bhp at the top end, if they could be found back at the low and mid-range.

Both 500 and 750 used four-petal reeds and seven port cylinders, as had been developed on Yamaha's motocross machines. The seven ports consisted of inlet and exhaust ports, four transfer ports and a gulley or finger port cut from the inlet tract to the cylinder wall just above the inlet port. Cylinders on the 500 measured 54mm × 54mm and were fitted with 34mm carburettors, confirming the concept of it being a doubled-up 250.

By July 1972, a 500cc engine was running and in September the machine was wheeled onto the test track for chief test rider MotohashiSan to put it through its paces. Hideo Kanaya, slated to join Jarno Saarinen in the 1973 500 team, was testing the bike by November 1972, to be joined by Jarno in January 1973. The bike Jarno rode in Japan was not fitted with reed valve induction. It seemed quite literally a doubled up 250 and was very peaky. Perhaps responding to his comments about peakiness, the reed valves

The final version of the OW19 at Yamaha's test circuit at Iwata early in 1973.

were fitted and appeared on the bike for Saarinen's first European shakedown test at Zolder in Belgium. This was the first he knew of the change of engine induction system. There had most likely also been a management directive to publicize the reed valve street two-strokes, still Yamaha's bread and butter, through a reed-valved MV-beating 500.

Although rumours had been circulating towards the end of 1972, Yamaha were very tight lipped about the plans to campaign in the 500 class. They were paranoid that any useful information might leak out to MV, enabling them to prepare more fully for the defence of their title. In the end, in February 1973, Rod Gould confirmed that the official works team for 1973 would consist of Jarno Saarinen and Hideo Kanaya riding TZ250s and the four-cylinder OW19. Their support team was headed by MisaguchiSan, and Nobby Clarke and Vince French would look after the 250 machines of Kanaya and Saarinen respectively. The OW19s were entrusted to two Japanese mechanics, SaitoSan looking after Saarinen's machine and OsachanSan after that of Kanaya.

A month later, MV were talking of withdrawing from racing, claiming that a country-wide strike of metal workers had delayed the development of the new 500-4, which they had been preparing to meet the expected Yamaha challenge. This pre-season press release might just come in useful if the Yamaha challenge succeeded. As it was, they started the season with the old three-cylinder four-valve engine slotted into a massive cradle frame with drum brakes front and rear and weighing in at a portly 320lb (145kg). Their chances did not look good.

The race debut of the OW19 was to be at a warm and sunny Paul Ricard circuit for the French GP. It was a fairy-tale debut with Saarinen leading off the line and maintaining a 20-metre advantage over Agostini until the Italian crashed on the ninth lap while trying to hang on. Phil Read had elected to race the new engine in the old chassis and was having a ding-dong battle with Kanaya for second place, which he eventually managed to secure, leading Kanaya over the line by two seconds. The team were happy but did not let it go to their heads. At the end of the day they rushed off to the Salzbürgring which they had hired for a few days of testing prior to the GP on 6 May. If the warm

The Crown Jewels

Saarinen hurries past the Armco at a wet and cold Salzbürgring in Austria on his way to 500 win number 2.

The West German GP at Hockenheim could have been victory number three for Saarinen, but for a broken chain and a hard riding Phil Read.

The Crown Jewels

and sunny French GP was good the cold and wet Austrian GP was better. Saarinen won again and Kanaya was second, after both of the MVs had stopped with engine problems. No-one else finished on the same lap. The 250 had an identical result, Saarinen first, Kanaya second, no-one else on the same lap. In fact this was to be one of Yamaha's most successful GPs of all time, as they won all the day's solo classes. A week later it was Hockenheim, but the winning streak turned out to be too good to be true. Kanaya retired with crankshaft failure and Jarno's chain broke on the fifteenth lap as he accelerated out of the Ostkurve at the back of the circuit. Until his retirement, Jarno had been lapping a few metres behind Read, biding his time, studying the lines, waiting for the mistake, confident he could pass as he pleased. Ago's 500, actually a 350 bored out to 430cc, dropped a valve and he retired. So after three races Saarinen was leading Read by three points with nine races to go. Next stop was Monza and an appointment with a slick Curva Grande, Passolini's sliding Harley-Davidson and oblivion. The deaths of Passolini and Saarinen, both of outstanding natural talent, were met with shock and grief throughout the sport. There is little doubt that Saarinen would have gone on to be World Champion in the 500cc class and would have been a strong challenger to Sheene in his years of Suzuki domination during the mid-1970s. The books will record his single 1972 250 world title, but history will recognize him as one of the most exceptional road racers of all time.

The official works team was withdrawn for the rest of the season as a mark of respect for Saarinen. The OW19s were returned to Japan to await the 1974 season. Although deeply upset by the death of Saarinen, Yamaha could not postpone indefinitely their challenge for the 500 title. Their search for a new rider coincided with Agostini's own frustration with the development cycle for MV machines and growing friction between himself and team-mate Phil Read. After much cloak-and-dagger stuff with Rod Gould, Yamaha's European PR

The first European appearance of the OW20 with monoshock rear suspension was at Modena in March 1974.

The Crown Jewels

chief, Ago announced his move to Yamaha at a press conference in December 1973. The stage was set for his titanic two-year struggle with Phil Read on the MV.

OW20

Tepi Lansivouri also joined the Yamaha team for 1974, initially as the number one rider, but ending up playing second fiddle to Agostini when the latter's late signing was announced. Agostini travelled directly to Japan in December for a few days and tested the OW19's successor, the OW20. Although told of its existence during the contract negotiations, this was his first exposure to the monoshock 500 that had been developed after the single shock YZ250 motocross machines that had won the 250 title for Yamaha. Ago revelled in the attention he received on his visit after the neglect and friction of his last MV season. Working with the design team engineers, changes were made to the prototype chassis, the engine was moved forward, the swing-arm was lengthened, the De Carbon rear shock was fine-tuned and some new gear ratios were chosen for the six-speed gearbox. The Yamaha crew listened to Ago's suggestions, clarified his requirements where necessary and implemented them with the legendary Japanese speed. Back in Japan just before the season start, six OW20s were lined up for him to test and make a final selection of two machines for Europe. This was superstar treatment and it felt good.

Initially, the Japanese needed to be convinced of Agostini's analytical skill, as Mac Mackay, Ago's 350 mechanic at the time, recalls.

> Agostini was very good at setting up a bike, as the Japanese soon came to understand. At Modena for the first 350 tests in March 1974, Ago came into the pits and said, 'Move the rear wheel back 5mm' before taking the other bike out on the circuit. Misoguchi, the team leader, says to me 'Don't change it, don't change it...'. Ago takes the bike out on the circuit again and pulls in after two laps and asks me 'What have you done to the bike? I told you to move the rear wheel back 5mm. What did you do?' I answer, 'Don't ask me, ask him...'

In the paddock at the 1974 Dutch GP, Agostini's OW20 awaits the start of its last GP, which it won.

It was to be a season of mixed fortunes for the new monoshock 500. It made an inauspicious start at Clermont-Ferrand where a gearbox bearing broke and forced Agostini's retirement while he had a comfortable eight-second lead over Phil Read on the MV. At Imola he ran out of fuel as he started the last lap narrowly ahead of Bonera's MV, but wins at Salzburg and Assen kept his title chances alive. Tepi took some time to adapt to the new machine and spent the whole season in Ago's shadow. Only Agostini was provided with new material throughout the season, and he also had first call on the pool of spares available to the team. Vince French, looking after Tepi's bikes was sometimes barely able to assemble a competitive machine due to unavailability of spares.

OW23

The OW20 engine was essentially unchanged from the OW19 ridden by Kanaya and Saarinen in 1973. By Spa in July 1974, a new engine had been prepared and was shipped to Belgium for Agostini's scrutiny and use if he preferred it to the OW20. It was the OW23, essentially an OW20 that had been on a drastic weight-reduction programme. Weighing in at 292lb (132kg) it was 46lb (21kg) lighter than the bike Tepi was to continue to run all season. A re-design of crankshafts and transmission had reduced the engine width by 20mm and length by 30mm and magnesium crankcases were fitted. Worrying about a repeat of the Imola fiasco, an extra fuel tank was fitted in the 'aerodynamically designed' seat, which extended down into a fairing around the monoshock swinging arm. Adjustment of the ignition at the start line, after the warm-up lap had shown the bike to be slow, did not fix the problem, as Phil Read ran away from Agostini in the race to win by more than a minute. The OW23 was to be raced one more time in 1974 at Anderstop, but a seizure of Barry Sheene's new RG500 Suzuki on the first lap took Agostini out and into hospital with a broken shoulder. Tepi went on to win

The first appearance of the slimline OW23 at the 1974 Belgian GP.

The Crown Jewels

One of the classic races of the 1970s, the 1974 Belgian GP. Agostini (4) with the OW23 was expected to beat Read (2) on the MV. Instead Read shattered the lap record and beat Ago by more than a minute.

in Sweden, but Phil Read cleaned up in Finland and Czechoslovakia and kept the title in Italy.

It was a big disappointment to both Agostini and Yamaha, but there had been exceptional circumstances and it was clear that both the OWs and the MVs were capable on their day of winning GPs. It was some consolation to the factory that they took home the manufacturers' title, but it was the riders' title that really counted. The teams had worked intensely to achieve the title and had provided very effective rider support. But they were only effective when everyone understood their role within the team. It took the Western mechanics a little time to fit into the Japanese teams as Mac Mackay explains.

> The mechanics were asked to provide information on the machines' performance and reliability, but were not expected to question the decisions behind the work they were asked to do. Misaguchi used to get angry with me and shout 'Why you say why', and I would explain that I wanted to understand what was happening, but that was not accepted as being something we should need to know. He would say 'Your responsibility is very small, mine is very big. We'll do it my way.' Despite this, Misaguchi was an excellent, charismatic team manager, always there through the long nights, ready to pitch in whenever required.

Yamaha's advantage was that the OWs were in their second year of development, whilst the MVs could be developed only marginally further. Time was on Yamaha's side.

During 1974 and 1975, the OWs were not the only four-cylinder 500s being run in the GPs. At the start of the 1974 season John Dodds and Dieter Braun had entered TZ700 machines with 250 cylinders fitted. They had both found the resulting bikes far from easy to ride and slower than the bored out 350s run by most privateers. With the TZ700 engine becoming increasingly popular in the sidecar classes in 1975, enough expertise on the correct way of performing the conversion had been gathered by the

The Crown Jewels

1975 season to make Jack Findlay's TZ700 500 moderately successful, with third places in Belgium and Finland to prove it. The biggest problem with the engine was its tremendous thirst, causing Jack to run out of fuel twice, and necessitating a pit stop in Belgium to tank up.

500 DEVELOPMENT – OW26 AND OW29

Detailed improvements were made to the OW23 during the winter of 1974–75 but externally the bike appeared unchanged. Fine tuning of carburation, ignition timing, exhaust design and port timing combined to reduce the massive thirst of the 1974 engine as well providing a slight power increase. Life had been made easier for the mechanics by the adoption of a side-loader gearbox. The art of the correct selection of internal gear ratios had become increasingly important with the 500. The OW19's power band had been less than 2,000rpm wide and it was becoming narrower. During the 1974 season, the need to split the crankcases to change the internal ratios had hampered the selection of the best combination. Now, having removed the clutch from the right-hand end of the transmission mainshaft, the removal of a further four bolts would enable the complete gear cluster to be drawn out of the crankcase.

The improvements to the 500, now designated the OW26, were enough to finally beat the MV in its ultimate development form. Phil Read rode the wheels off the Italian machine, forcing the title once again to be decided at the end-of-season Czechoslovakian GP. Despite Read's win, Ago's second place was enough to give him the title by eight points. In the six races that Ago finished, he won four and came second twice. The three failures to finish were caused by seizures in Austria and Belgium and a flat front tyre in Sweden. It had been a monumental struggle, for both man and machine, but both Yamaha and Agostini had triumphed.

The official portrait of the 1975 OW26 which Agostini used to win his last and Yamaha's first 500 class title.

The Crown Jewels

A mid-season shot of the OW26 at the 1975 Belgian GP. Agostini was forced to retire from the race when water leaked into the engine.

The 1975 season had been characterized not only by the magnificent struggle between MV and Yamaha, but also by the increasing competitiveness of the new Suzuki RG500s, which Barry Sheene had used to win the Dutch and Swedish GPs. It was clear that these machines were the new challenge for Yamaha, who decided that a year's sabbatical from the world of GP racing was needed to design and build a new piston-port engine to challenge the rotary-valved square fours. Johnny Cecotto managed to prise Ago's machine (now designated the OW29) out of Yamaha's hands for the 1976 season, but it turned out to be a disaster, despite an encouraging initial second place at the opening French GP. The Suzukis, now available to privateers, were clearly faster and Johnny ended up crashing so often trying to run with them that the team had run out of spares by the time of Dutch TT. Sensibly, Cecotto decided not to attempt the impossible and concentrated on the 350 and 750 classes only for the rest of the 1976 season.

OW35

At the start of 1977, a new machine was available for the Yamaha factory team of Johnny Cecotto and Steve Baker. The OW35 might have appeared to be an evolutionary development of the OW29, but in fact the blood-line originating from the OW19 was broken by this machine. This was really the Mark II version of the piston-ported 500 factory machines. The most noticeable external change to the engine was the disappearance of the reed valve induction. Cecotto had not been able to run with the RG500s at the start of 1976, so Yamaha decided that the reed valves, with their inherent barrier to full inlet flow, must go. Bell mouths were fitted to

Steve Baker's first GP season brought him the runner-up spot in the 1977 championships. His machine, the OW35, had lost its reeds and marked the start of the second generation of the piston-ported 500s.

the carburettors and Yamaha relied on the tuned induction-tract length to support the inlet port timing. Powerjet valves were fitted to the 34mm Mikuni carburettors to improve the top-end power. As on the original TZ750s, there had scarcely been enough room under the engine for the exhausts of the correct dimensions. Now, with a slight change in frame design, it was possible to thread the outer left-hand exhaust round behind the engine to poke out under the seat on the right-hand side. The pipes were now capped with mufflers to meet the stringent noise restrictions imposed by the FIM during 1976. The engine itself changed from the square 54×54mm used since 1973 to a short-stroke 56×50.6mm This was provoked solely by the desire to increase the top-end power of the engine. The shorter stroke necessitated smaller, and hence lighter, flywheels that had the secondary advantage of improving crankshaft reliability.

The engine now ran backwards, i.e. the cranks turned anti-clockwise. Yamaha intended to open up the inlet port as far as possible to flow the maximum amount of fresh charge. With a forward running engine, this would reduce piston reliability enormously, as the skirts would be inadequately supported during the power stroke and would quickly fracture. With the engine running backwards, the skirts would be supported by the solid cylinder wall under the exhaust port and could be expected to live long and happily in the engine. Now that the engine was running backwards, an extra jackshaft needed to be placed between the cranks and the transmission to correct the direction of motion before it reached the rear wheel. A vertical shaft driven by the jackshaft was used to power the water-pump and gearbox lubrication pump.

The frame design was essentially unchanged, but there was an attempt made to provide better aerodynamic form to the machine. From 1974 to 1977 most seat designs had included a slight 'spoiler' effect to the glass-fibre behind the cushioning. At Spa in 1974, Agostini had fitted a fuel tank into a seat that had differed by tapering from the width of the seat to form a wedge whose apex was at a point just above the rear wheel. It was a seat of this shape that was adopted first by Yamaha and within a couple of GPs, Suzuki followed suit. It is doubtful if either the previous 'spoiler' or the new 'teardrop' form had much influence on either down-force generation or drag coefficient reduction.

Yamaha had high hopes for the 1977 season. They had a machine that was now producing 110bhp, although admittedly over a painfully thin power band from 10,000 to 11,000rpm. Nevertheless, in the hands of Johnny Cecotto, Giacomo Agostini and US superstar Steve Baker, a trio of highly gifted racers of proven ability, they expected to take the 1977 title. It was to turn out to be a painfully disappointing season for both riders and factory. In a straight fight with Suzuki number one rider Barry Sheene, the OW35 came off second best every time. Cecotto managed to win two GPs, in Finland, where Sheene's machine overheated, and in Czechoslovakia which Sheene did not bother attending, as he already had the Championship tied up.

Baker managed two second places and three thirds and Ago, in his last GP season, managed two second places. But Yamaha did not want rostrum places, they wanted wins and the crown. Baker was sacked, despite taking the F750 title, while Ago was refused machines for 1978 and finally hung up his leathers. Cecotto's two wins were enough to enable him to stay on for the 1978 season.

What had gone wrong? Once again it was a case of dyno blindness. Yamaha had managed to produce a machine with more top end than the Suzukis and they were faster, but the razor sharp power band combined with a flawed chassis meant that the Suzukis outrode them round the corners. During his winter trip to Japan at the end of 1977, Roberts was able to ride Baker's bike and he concluded that the swing-arm pivot was too high and that the monoshock rebound damping was too severe. With his superb analytical skills Roberts was able to identify and ease these problems to improve the 1978 machine enormously.

Yamaha were fixated by engine design at

It's hot and dusty at the 1978 Venezuela GP as Kel Caruthers prepares the 1978 OW35 for the start of the GP season. Bad fuel killed all OWs in the race.

the time with all their energy going into perfecting the piston-port design. They, more than anyone, were aware of its inherent restrictions, but, as in the application of the reed valve, they were determined to push back the frontiers of knowledge in their quest for success on the race track. The reed valve had been the 'solution' to the asymmetric timing requirements for the inlet port; now it was time to see what could be done to optimize exhaust-port design. The first step in this direction came at the 1977 Dutch TT, when the Yamaha mechanics suddenly started covering the bike with tarpaulins as soon as it returned to the pits. Despite persistent attempts to photograph the bike without fairing during the rest of the season, no-one was successful. A number of rumours started circulating, that Yamaha were using fuel injection, had adopted some radical steering system, a feigned attempt to put the wind up Suzuki, and the use of four separate cylinders in place of the pair of twin cylinder castings that had been used since 1973. If that photograph had been taken, it would indeed have shown four separate cylinders, in itself not particularly spectacular, perhaps facilitating transfer port design, but nothing special. In fact Yamaha needed to test these cylinders in preparation for their new 'gizmo', the exhaust powervalve.

Exhaust port design is just as complex as inlet and transfer design, with its own set of conflicting requirements. The exhaust port should be timed to open far enough ahead of the transfer ports to ensure that the pressure in the combustion chamber has dropped to the pressure level within the crankcase. If the pressure has not dropped sufficiently, the exhaust gases will hinder the smooth flow of the fresh charge into the combustion chamber. This will certainly reduce the engine's combustion efficiency. At high engine speeds, the speed of the piston is so great that the exhaust port needs to open much earlier than it does at low engine speeds in order to guarantee adequate drop in pressure. On the other side of the equation is the need to get the exhaust port closed again before any of the precious charge flowing from the transfers is lost into the exhaust port. Exhaust pipe design helps enormously with recovering lost charge, but at low engine speeds charge will still be lost if the engine has an exhaust port high enough to work well at the top end. Hence the dichotomy; the optimum engine should have variable-height exhaust ports to work well at low and high engine speeds.

Mechanisms involving the physical movement of the cylinder wall were quickly rejected as causing more problems than solutions. Experimentation showed that it was not necessary to have a perfect gas-tight fit between the piston and the cylinder to make the engine believe its port height had changed. Yamaha showed that the required effect could be obtained if a small cylindrical valve was inserted above, and partially protruding into, the exhaust port tract, close to the cylinder wall. By cutting away part of the valve face and rotating the valve it was possible to vary the amount by which it protruded into the exhaust tract. It was effectively lowering the height of the exhaust port. As engine speed rose and the valve rotated, less of the valve projected into the exhaust tract until, approaching top engine speed, the valve was flush with the tract roof. The powervalve had been born.

Separate cylinders were necessary to mount the valve on pivots in each of the four castings. All the valves were joined to their neighbour to enable them to turn as a single unit. The drive for the rotation of the valve was provided by a tiny electric motor that was housed in the fairing next to the tachometer. The attitude of the valve (open, closed or in between) was compared with the engine speed displayed by the electric tachometer and the valve position adjusted

The Crown Jewels

to a pre-programmed value. This all happened many times a second to keep the valve accurately positioned.

It was expected that this would cure at least some of the problems that the OW35 had experienced in 1977 with the lack of mid-range power in comparison with the Suzukis, and indeed it did. The Yamaha 'team' for 1978 was to be Johnny Cecotto, Kenny Roberts and Takazumi Katayama. In fact they were three separate teams that all happened to be riding works YZR500 Yamahas (still identified as OW35s). Katayama had a European-backed team, Cecotto the South American and Kenny Roberts the USA team. The first GP of the year was at Venezuela and naturally enough Cecotto got the only 500 engine with power-valves for his home GP. Despite Cecotto's fastest practice time, the race turned out to be a disaster, with all three Yamaha bikes retiring. Cecotto had tyre problems, but the engine had been fine. By the first European round in Spain all three riders had the new engines and the battle with Suzuki could begin in earnest.

Having set pole position time in practice, Roberts quickly moved to the top of the field and pulled away from the leading Suzuki of Pat Hennen. With just eight laps to go and a comfortable 12-second lead, the slides on two of the carburettors stuck open. It took all Roberts' skill to coax the bike home, but he could not prevent Hennen passing him to win his third GP. A bitter-sweet result for Roberts, the Yamahas having given the impression of having (most of) the Suzukis beaten for top speed, but frustration at the carburettor problem. Root cause analysis revealed that the aluminium slides of the magnesium Mikuni carburettors had a tendency to burr and stick. From then on steel slides were fitted and the problem did not return.

On to Austria and the ultra-fast Salzbürgring proved that the Yamahas were a lot faster than the Suzukis. Sheene complained of the '20mph top speed advantage of the Yamahas', a vast overstatement, but it must have seemed like that as the cracks appeared in his two-year domination of the 500 class. Now the Roberts roll started, with wins in Austria, France, and Italy, a second to Cecotto in Holland, and a second to Hartog's Suzuki in Belgium. In Sweden the strains of the season caught up with him and he crashed in practice while testing his Goodyear tyres. Badly concussed he started the GP anyway and somehow managed to finish seventh, despite double vision. Early season challenger Hennen had crashed at Bishops Court during the Isle of Man TT and was to remain in a coma for several months, tragically terminating a career in road racing that seemed only a step away from big time success. Sheene's win in Sweden helped him claw back some of his points deficit, but machine failure for both riders in Finland, and Robert's win at a chaotic wet British GP, gave the latter an eight-point lead before the German GP at the infamous old Nürburgring. Mercifully dry throughout practice and race day, Kenny learnt the complex 14 mile (23km) circuit well enough to grab second fastest practice time behind Cecotto. With Cecotto well out of the title chase, Kenny might have hoped for some cover from him during his own challenge, but Cecotto took off like a scalded cat with Virginio Ferrari, who was getting his first works Suzuki ride, following close behind. Roberts caught the pair and slotted in behind Cecotto, only to find his fellow Yamaha rider weaving in an attempt to break the tow. Roberts let the pair go and concentrated on staying ahead of Sheene who was just behind him in fourth place. With a bike set up for a finish he might have been pushed to keep Sheene behind him, but the challenge never materialized and he came home in third place to collect his first 500cc World Championship title.

All three of the OW35s led the 1978 Dutch TT at some point. Here Katayama and Roberts battle it out but in the end Cecotto was the winner.

The statistics make it look a relatively easy win, but the bike was far from perfect. The engine was good in that it delivered well in both mid-range and at the top end, and Roberts in particular grew to appreciate what he called the 'pushing power' of the piston-ported OWs. The guts of the power band, wheel-spinning territory, were from 10,000rpm to 12,000rpm, although it would pull from 6,000rpm. The biggest problem was the rear suspension that would deteriorate significantly by the last third of each race, preventing a consistent race-long strategy. The team attempted many different ways to prolong the rear shock effectiveness through the race, but none really worked.

OW45

The feedback that Roberts had passed back to Japan via the team conduit engineer and translator, Maekawa, as well as that coming from Cecotto and Katayama, convinced Yamaha to address the suspension problem for the 1979 machines. Their solution was the adoption of a De Carbon rear suspension unit with both coil spring and nitrogen pres-

The Crown Jewels

The 1979 TZR500 was coded the OW45, and Cecotto and Roberts were each entrusted with a pair of the machines. Only minor suspension changes distinguished it from the OW35 of 1978.

sure acting as a secondary spring. Little else of significance changed on the new YZR500, coded the OW45.

Cecotto and Roberts were to form the Yamaha challenge for 1979, once again riding for different teams. As in 1978, the season did not get off to a good start. During January testing in Japan, Roberts was trying out new tyres when he lost the front end in a fast corner and hit the crash barrier at 90mph, breaking his back and foot and rupturing his spleen. He was nowhere near recovered by the mid-March Venezuelan GP and had to give it a miss. Cecotto followed the tradition of retirement from his home GP. At the Austrian GP at the end of April, Cecotto crashed and splintered his kneecap, causing him to miss the first half of the season. The weight of responsibility for Yamaha's 500 GP campaign rested solely on the aching back of Kenny Roberts. He rose to the challenge incredibly, winning in Austria, coming second in Germany, and winning in Italy, Spain and Yugoslavia. As in 1978, Roberts seemed to have excellent results in the first half of the season enabling him to establish a substantial points lead before the European riders hit form. Virginio Ferrari had not let the gap open that year and was only six points behind as they went to the Dutch TT at Assen. Ferrari won and Roberts could do no better than eighth with a rear shock that had lost almost all of its damping. This had been caused by a leaking bladder seal that had allowed some of the nitrogen to mix with the oil. Onto Scandinavia, where Roberts again had rear suspension problems in Sweden, restricting him to fourth, and crashed in Finland to keep him back in sixth.

Fortunately for him and Yamaha, Ferrari's trip was even more disastrous and the Continental circus came to the UK with Roberts showing a seven-point lead. Silverstone was to be the scene of the famous race-long duel between Sheene and Roberts, with Roberts taking the flag by half a machine's length. His third place in France three weeks later gave him his second world title and the satisfaction of knowing he was still the king.

OW48

Yamaha were of course delighted with their second consecutive title, but Roberts was not so sure that his piston-ported straight-four would be able to fend off the Suzukis for much longer. His early season speed advantage had been lost by the end of year as shown by the ding-dong battle with Sheene at Silverstone, one of the fastest GP circuits. Roberts was pestering the factory for another 10 to 15bhp, and the factory kept telling him that it was not possible with the current configuration. Their alternative was to reduce the weight of the machine, although both short and long-term projects were being worked on for more power. The OW45 had not used a lot of aluminium chassis parts, although the exceptions were the fork yokes and the triangulated box-section rear swing-arm. The OW48 was going to have to lose a substantial amount of weight and that would be achieved by replacing the steel cradle frame with an aluminium version. The basic design of the frame was unchanged, although Roberts requested a slight lowering of the engine and a fraction more weight on the front wheel. The only significant change in the engine was a move away from the cylindrical powervalve to a guillotine version that was more compact and considered to be slightly more efficient, as it was possible to locate it closer to the exhaust port. Carburation was another area that came under close scrutiny with a variety of Mikuni magnesium and aluminium

The official portrait of the 1980 OW48 shows a steel-framed bike, although black paint on the aluminium frame made it hard to tell the difference from a distance.

The Crown Jewels

units available, and 34, 36 and 38mm carburettors with both cylindrical and flat-slides were used throughout the year. The flat-slide units worked best at high-speed circuits, whereas the cylindricals produced better out-of-corner torque on the twisty tracks.

Roberts started the season on the OW48 with the choice of either the steel or aluminium frame but both with the slight re-positioning of the engine. Yamaha, never one to attract attention to machine developments, craftily painted the frame black, so as not to draw attention to the change of frame pipe metal. Initially the aluminium frames proved to be too flexible and as the season progressed, double box-section tubing was used on the upper rails of the cradle to enhance stiffness. With the Venezuelan GP no longer on the calendar, Roberts

Kenny Roberts gets the OW48 fully grounded on his way to win at the 1980 Spanish GP.

The reverse cylinder OW48R was Yamaha's last gasp attempt at keeping the piston-ported engine competitive. It didn't help much.

Setting the record that stood for twenty-one years, Bill Ivy flies over Ballaugh Bridge at the 1968 125 Lightweight TT. There is no truth to the story that he stopped for a smoke on the last lap to ensure that Read would win as team orders dictated.

Complex scoring of points denied Read the 1967 250 title. Despite having more points he had fewer victories so Hailwood took title. Here at the East German GP at the Sachsenring, Read was unbeatable.

Read at Braddan Bridge during the 250 1968 IoM TT. A puncture took him out of the race, but Ivy was really fired up and would have won anyway, as team orders dictated.

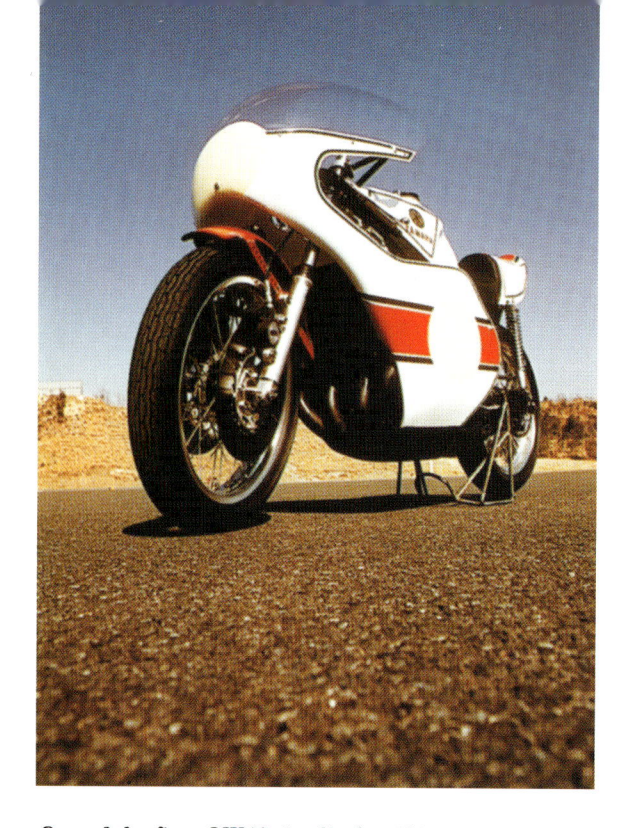

One of the first OW19 4-cylinder 500 prototypes developed to take the 1973 title from MV Agusta. It would have happened but for Saarinen's tragic accident.

How to make a motorcycle racer happy. Give him a works contract to ride the cutting edge of two-stroke technology against a 4-stroke relic of a bygone era. Hideo Kanaya was a happy man when he first got to see the bike at Yamaha's Iwata test track.

Saarinen acknowledges the cheers of the French crowd for a historic moment in motorcycle racing history; the first 500cc win by a two-stroke machine.

The Japanese mechanics seem to think that Vince French and Risto Tennhunin, Lansivouri's mechanics, have the preparation of the OW20 for the 1974 Belgian GP at Spa under control.

Saarinen rode the OW19 to a victorious debut at the warm 1973 French GP at Paul Ricard. Ago's MV expired trying to keep up and Read just managed to hold off Kanaya for second place. MV were in trouble.

The times they are a-changin'. Ago wearing a classic pudding basin helmet aboard the classic 350 class winner, the 3-cylinder MV, had his hands full with Phil Read on his TR2B at the 1971 Dutch TT. Ago won the race but the writing was on the wall ...

The US Yamaha team hit Europe with the OW31, OW35 and this very business-like TZ250E slotted into a scaled-down replica of the OW31 frame. A year later it was available to all as the TZ250F. Here mechanic Trevor Tilbury preps the bike for the Dutch TT win.

Patient development by Kent Anderson of the air cooled street bike was rewarded with the 1973 world title. The 1974 bike shown here had changed little and retained the title, but a year later Morbidelli had arrived and was invincible.

Sizing up the track and the bike. Agostini gets down to some serious training in preparation for the 1974 Daytona 200. A couple of days later he had taken the TZ750 to an impressive win on his debut race for Yamaha.

In-journey entertainment for the passengers of the Trans-Finland express in the form of the 1979 Finnish GP. The infamous railway crossing at the equally infamous Imatra circuit. A couple of years later the anachronism was gone.

The official factory portait of the 1980 OW48 shows it with the old round-tubed steel frame. Almost immediately it was replaced with a square-section aluminium design craftly disguised by black paint. No one was fooled and it was not so effective.

What an underrated rider Graeme Crosby was. Here he takes the ageing YZR750 to one of its last victories at a major International race, the 1982 Daytona 200. Runner-up in the 500 World Championship, he turned his back on the GP world, sick of the politics and subterfuge.

The 1982 TZ500J was a replica of the OW48R with reversed outer cylinders, which Roberts had used occasionally two years previously. It was expensive, unreliable and not competitive. Not one of Yamaha's success stories.

During the first half of the 1980s, the TZ was the mainstay of 250cc racing throughout the world. This was the 1983 K model, an expensive but only slightly better version of the TZ250H, the first Powervalve model that had appeared two years before.

The last of Kenny Roberts' warhorses was the 1983 OW70, which served him well in his epic battle with Freddie Spencer. Chassis changes during the year and the nascent involvement of Öhlins almost snatched the championship from the jaws of defeat.

He practically had the 1984 title sown up when he arrived at the Belgian GP at Spa, but that was no reason to take it easy. Lawson struggles to keep the front wheel of his OW76 on the ground on the exit from La Source.

In colours that were later to become associated with Honda, Carlos Lavado rode the new YZR250 to its first and his second world championship. The character of the man meant that he either fell or won.

Showing the guts and determination that would bring him three titles, Wayne Rainey squeezed past Lawson on the last lap of the 1989 German GP. He won the battle, but in the end lost the war that was the 1989 GP season.

As far as he was concerned, John Kocinski was fighting Luca Cadalora for the 250 crown in 1990. In the end Cadalora came third, with Cardus proving a mild threat to Kocinski's domination of the class that year.

Battered and bruised from a practice spill, Wayne Rainey holds off Luca Cadalora at the 1993 British GP. In the end Cadalora passed his teammate to take the win. A few weeks later Rainey's racing career had ended.

Starting the season well and getting better Tetsuya Harada took the 1993 250 GPs by storm. He was the only person able to ride the TZM250 competitively and the world title was the consequence.

Wayne Rainey showed his personal qualities when he refused to let pain and despair destroy his life. We will never forget the demonstration of his outstanding riding qualities on the red and white Marlboro Yamahas.

The Crown Jewels

Boet van Dulmen got semi-official rider status with the 1981 TZ500J featuring reverse-cylinders. He managed a couple of second places during the season.

it was intended for the US market; it was Europe where it was expected to sell well. This would not have been too bad if it had been a good motorcycle, but it wasn't. It was underpowered and steered poorly, as should have been reported by the Japanese test rider Takai after his eighth place at Sweden in 1979 on a prototype of the TZ. There was too much flex in the frame where the narrow rear of the engine hung in extension tubes from the cradle. The front forks were not stiff enough. The engine was a good 20bhp down on the OWs, barely in the same ball park as the RG500 Suzuki, the machine against which it would be bench-marked. Fiddling around with the carburation helped some, American Rich Schlachter for instance using Lectrons with some success.

The 1980 GP season was a bitter disappointment to the TZ500 owners. The only real success came at the damp Dutch TT which Jack Middelburg won, but he was riding a TZ500 engine in a Nico Bakker frame with RG500 front forks, and he was unable to come near to this performance for the rest of the season. For 1981, the TZ500H was available with detail changes only to the specifications. Most significant of these were

The Crown Jewels

suspension changes front and rear. The rear damper was re-jetted and was fitted with a new spring. The front end was completely re-worked resulting in new jetting, springs and stanchions. The changes did not help much; the RG500 remained the privateer's first choice.

In an attempt to provide Roberts with some support for the season, Boet van Dulmen, Michel Futschi, Sadao Asami, Christian Sarron, Marc Fontan and Barry Sheene were provided with OW48Rs in aluminium frames. Sheene received his new OW54 quite early in the season, but van Dulmen and Fontan had some good sub-rostrum places topped by van Dulmen's second place at the wet Dutch TT.

The performance of the OW48Rs had been good enough for Yamaha to bring out yet another TZ500 for 1982, the TZ500J. This was an OW48R replica with steel rather than aluminium frame and mechanical powervalve. Only Steve Parish and Marc Fontan rode these machines in 1982, but with little success, and with the bad reputation of the G still intact, there were few sales. The engine was to live on for many years to become the mainstay of the sidecar racers, who found the 'pushing power' of the engine ideal for their outfits. Even today, eleven years after production ceased, many of the sidecar engines use either original or clones of the Yamaha TZ500 crankcases.

The piston-ported four-cylinder had brought Yamaha four championship titles in the six years it had been supported by the factory. It had been the guinea pig on which Yamaha's ideas about transfer shapes, inlet and exhaust timing had been tested and improved. That it should be put aside as new ideas were developed was as certain as that night should follow day. Evolution can go just so far; sometimes revolution is the order of the day. It seems, however, to be hardly coincidental that three of motorcycle racing's greatest names rode and won on the four-cylinder machine: Jarno Saarinen, Giacomo Agostini and Kenny Roberts. As surely as their names will not be forgotten in the passage of time, neither will the contribution of the first 500cc two-stroke world championship machine.

7 The Beast
TZ750 1973–9

TZ750 ORIGINS

The birth of the TZ750 can be traced back to the Sunday morning of 31 October 1970, when the seventeenth annual Tokyo show opened its doors for the first of six days. Two years earlier Honda had astonished the world with the première of the CB750, a milestone in the history of motorcycle design. In 1970 it was Suzuki's turn, for their stand was dominated by the shiniest, largest capacity, most complex two-stroke road machine ever built; the prototype of the GT750 water-cooled triple. It seemed to everyone that Suzuki had picked up the gauntlet to challenge Honda's four-stroke with an equally sophisticated two-stroke.

Yamaha were not happy. During the 1960s they had established a name for themselves as the pioneers of innovative two-stroke design for both road and racing machines. Their commitment to the use of the two-stroke for road machines had been compromised by their foresight in realizing the limited future that awaited it. Their first four-stroke had been announced at the 1969 Tokyo show in the form of the 650cc XS1. Far from being innovative, it was probably the last example of the renowned Japanese trait so common during the 1950s, of copying successful machines. It was a Triumph Bonneville clone. In the meantime, Suzuki had a 500cc twin in production and topped it with the GT750. Yamaha were being squeezed.

After the 1970 show had closed its doors, Yamaha management decreed that the 1971 show would feature an innovative Yamaha superbike and a design team was created to work on the project. The design team as picked from the small research amd development department. It was inevitable that the bike they were to produce would be a two-stroke as the department's expertise lay solely within this area. Sixteen years after the company's creation, the XS1 Bonneville clone was the only four-stroke in Yamaha's thirty bike line-up. The need to put one over on Suzuki led them to the 'superbike' of the 1971 Tokyo Show, the GL750. It was everything the GT750 had been, but with its four cylinders, double front disc brakes as well as water-cooling, it was clearly intended to put the GT750 in the shade. As icing on the cake, and to justify the 'innovative' stamp, both the newly introduced reed-valve induction and fuel injection were present. Yamaha stand crew and PR departments around the world were reluctant to enlarge on the standard bare-bones specification sheet that claimed 70bhp at the astonishingly low engine speed of 7,000rpm, running through a five-speed gearbox. No delivery dates could be given and questions on the price were met with a shrug of the shoulders. At the end of the show, the sole GL750 prototype returned to Hammamatsu never to be seen again.

While Yamaha continued their troubled struggle to establish a respected range of four-stroke road machines, the racing

Whatever happened to the GL750 prototype that is rumoured to have been transformed into the TZ750 behind the closed doors of Yamaha's research and development department?

department was not constrained by noise and emissions legislation being introduced in the USA. Their attention was being drawn to the increasing popularity of road-bike-based racing in the USA. Exemplified in Europe by the annual Daytona races that had been run for many years, an annual series of races had developed and was attracting both British and Japanese works support as well as increased interest from European riders. The air-cooled TR3s usually did not have the stamina to win the longer races but did perform well at other locations, particularly in the hands of Kel Carruthers. From 1972 both Suzuki and Kawasaki began competing in the AMA National Championships with 750 triples based on their roadbikes, indicating their determination to rule US road racing. Yamaha were not prepared to diverge from their plans to concentrate on four-strokes for the large capacity road machines. This seemed to exclude them from US and FIM F750 racing which was intended to be based on street-legal machinery. After lengthy discussions between Yamaha and the Japanese Federation, sub-contracted by the FIM to homologate machinery produced by the Japanese industry, it was agreed that a 750 four-cylinder two-stroke would be homologated, provided at least 200 were built. Essentially the homologation extended solely to the engine. This led the way to Hideto Eguchi announcing in August 1972 that Yamaha would be competing in the 1973 US and European 750 race series.

It turned out that he spoke rather prematurely, as the factory had their hands full with the 500cc GP four-cylinder they were preparing for Saarinen as well as the newly introduced water-cooled TZ350. Despite winning Daytona on one of these agile giant-killers, Kel Carruthers confirmed that it would be the full 750 they would be running in 1974. Also, not missing an opportunity to put the fear of God into the opposition, he reported tests on the 750 in Japan achieving a top speed of 183mph (295km/h). Then one day in June 1973, the phone rang in Kel's workshop in San Diego. It was Japan. 'Kel-san', they said, 'its ready for you. Come and see for yourself'.

The Beast

Beauty and the Beast at the 1973 Tokyo Motorcycle show.

Kel came, saw and was conquered. There was no doubt in his mind that Yamaha had a winner. He spent four days testing prototype 001 of the TZ750. The bike turned out to be familiar territory, in the form of a doubled-up TZ350, with some changes intended to *detune*, what would otherwise have been an unridable monster. Although appearing to be a doubled TZ, only a few parts were used from the 350. The engine was dominated by the set of two water-cooled cylinder pairs and the enormous radiator feeding them coolant. The cylinder heads were very similar to those of the TZ350, with the exception of a squish-band clearance dropping from 2mm to 1mm. As on the 350, each cylinder had a bore and stroke of 64 × 54mm, resulting in a total displacement of 694cc. The single most conspicuous difference between the cylinders of the two machines was the use of four-petal reed-valve induction on the 750, primarily to flatten and broaden the power delivery. As on all of Yamaha reed valve equipped bikes, a fifth transfer port was present on the inlet side of the cylinder. This took the form of a gulley extending from the inlet port up towards the cylinder head. This enabled the exhaust pressure wave to crack open the reed during scavenging and pull in some fresh charge directly from the inlet manifold, rather than from the crankcase. The result was a general improvement in scavenging waste gases from the murky cylinder wall furthest from the exhaust port. As with other Yamaha bikes using reed valves, both motocross and street, windows were cut in the inlet side of the piston skirt. The piston modification was intended to extend the inlet duration, i.e. enable the crankcase to be filled more completely. A few of the first factory 750s were delivered to the USA without this modification, presumably as it was felt unnecessary to tune the powerful engine at the risk of compromising the reliability of the piston. On the exhaust side, the port was also 1.5mm lower than that of the 350, reducing the exhaust-open duration, tending to

103

The Beast

The TZ750 was hyped up by the press into something far more than it really was. Fast, tractable, it became the cornerstone of large capacity racing throughout the world.

improve mid-range at the expense of top-end power. Transfer port timing was identical.

At the bottom of the engine, two crankshafts were located in the horizontally split magnesium alloy crankcases. The two crankshafts were not connected directly but both had 10mm wide gears mounted on their inboard end, driving a 20mm wide gear on the countershaft. The con-rod assembly was vintage TR3, but the cranks themselves were special for the 750. Each crank had the crankpins 180 degrees apart, resulting in cylinders 1 and 4 and 2 and 3 firing simultaneously. The countershaft was used to drive the clutch via the primary gear, as well as driving the water-pump, magneto rotor shaft and trochoidal transmission oil pump.

The dry clutch was a massive affair, well up to handling the 90bhp Yamaha were conservatively quoting for the engine. All components of both the clutch and gearbox were over-engineered. Yamaha were well aware of the danger of the gearbox failing at the 180mph (290km/h) speeds they had been working towards. The clutch contained seven plates of 152mm diameter and fed power into a six-speed gearbox. With the low-end guts of the TZ, there was no need for a low first gear and it was good for 70mph (110km/h).

The ignition system consisted of four coils powered by a capacitor discharge (CDI) magneto, driven off the countershaft. The CDI system had been blamed during 1973 for some of the weaker performances of the TZ350, so the 750 was given its own re-design. It proved to be well up to the job of providing sparks for the engine that ran to 10,500rpm. The exhausts were however a bit of a problem. There was just no room for them all if they were to be dimensioned something like the shape required of them by the engine. During the 1960s, as exhaust design became increasingly better understood, it was clear that the pipes should be as conical as possible. In this way, they would provide the correct timing of positive or negative pressure waves at the exhaust port as well as being most resistant to vibration caused by the wave travelling down the pipe. On the 750, it was not possible to keep all four pipes perfectly conical, as they swept down underneath the engine. Consequently, pipes for the middle two cylinders had

Simple but effective, the TZ750 was one of the best motorcycle engines Yamaha ever produced.

flat-sided mid-sections. Mikuni 34mm carburettors running a 300 main jet were fitted. The Autolube system was missing, presumed dead, for Yamaha finally acknowledged that no serious racer would leave the critical mix to the vagaries of an oil pump.

The frame was totally new, bearing little resemblance to other TZ frames. Initial prototypes of the 750 had been shoe-horned into larger scale versions of the TD/TR frame, but the bike had felt too heavy and cumbersome in the slower corners. The new frame was characterized by the upper rails dropping down to the rear engine mountings. Twin loops ran down from the headstock and under the engine to form a distinctive cradle. Bracing tubes were added between the upper rails, under the petrol tank and between the upper and lower rails just above the engine. Extensive gusseting was present around the headstock. The main tubes running under the engine were 28mm in diameter whilst the upper and sub-frame tubes were 25mm. The upper rails were spaced wide enough apart to allow the petrol tank to sink between them, supported on its outer edge. This helped lower the centre of gravity of the 370lb (168kg) ready-to-race machine. A heavily gusseted box-section swing-arm ran back to the two shocks that could only be adjusted for pre-load. The first prototypes had a drum on the back wheel, but that had made way for a single disc on the rear wheel and a pair of discs up front. Steering geometry was set up to help to make the bike as agile as possible with a rake of 26 degree and a trail of 3.25in (82.5cm). The bike was finished with the

The Beast

At Ontario in 1974, Gene Romero and Kenny Roberts check to see what Kel Caruthers has got for them this time. Looks like a pretty stock TZ750.

traditional Yamaha paintwork of white with a red stripe down the fuel tank and rear seat unit. This was the machine that was to achieve world-wide domination of 750 racing for the next eight years.

THE SUCCESS STORY BEGINS

The bikes that greeted Kel on his arrival in Japan were already well prepared. The only major problem that he encountered was a very bad high-speed weave above 160mph (260km/h). This was significantly improved by the addition of 5cm to the length of the swing arm and some adjustment to the rear shock damping. He also took the time to check the machines for maintainability and reliability bearing in mind that 200 were to be let loose on the world and not all would be falling into equally experienced and capable hands. He made a number of recommendations that Yamaha accepted despite the fact that some machines had already been assembled. These were despatched with a small kit of replacement parts to be installed by their happy new owners.

Kel returned to California, finished off the 1973 racing season and announced his retirement from racing in order to concentrate on machine preparation for the US Yamaha team. He denies this decision had anything to do with the 180mph (290km/h) bullet he rode those four days in Japan. By November the first two brothers of 001, 002 and 003, were at Ontario undergoing analysis by Gene Romero and Kenny Roberts and they were impressed by the bikes. The hype had made them out to be monsters in everyone's mind, but they were actually not that difficult to ride and to ride fast. During the test sessions, one of the problems that was to plague the first TZ750s reared its ugly head as one of the flat-sided exhaust pipes split. This was to occur with only too regular frequency and cost Kenny Roberts the win at Daytona, the first big international race for the TZ750 in March 1974. Fortunately, another bike prepared by Kel won the race, providing an ecstatic Agostini with a win on his debut race for Yamaha, his debut race on a two-stroke and his debut race in the USA. With something like fifty entrants for the race riding the bike, the results were packed with finishers on the 750, including all the places from twelfth to thirty-second. It had arrived and made sure everyone took note.

One group of people who were less sure that they wanted the TZ750 to stay were the technical sub-committee of the FIM. A couple of weeks after Daytona, they announced that the bike did not comply with the FIM

Although he didn't win Daytona in 1974, Roberts won most of the other US road races that year including Ontario.

regulations for F750 racing and that it was to be banned with immediate effect. Total uproar in the racing world ensued, it quickly becoming clear that typically there were two different versions of the FIM regulations in circulation, both pretty vague, but one talking about 'street legal sports bikes'. Despite universal criticism of both the ban and its timing, the ban remained in force for the duration of 1974, to be lifted for 1975 and subsequent seasons when less ambiguous regulations were issued.

Some meetings avoided the ban by withdrawing from the F750 series and staging an 'open class' race, permitting, even welcoming, the TZ. The first race to adopt this tactic was the Italian Imola 200, the Euro-Daytona. With many of the big names from the USA competing, plus the best European riders, the race was once again an Ago success story. After comfortably winning the first leg from Kenny Roberts, he held off a late challenge from Kenny in the second leg to win by almost 40 seconds total race time. Kenny had the satisfaction of setting a new lap record, 100,000 Italian fans had the satisfaction of Ago's win and Yamaha had the satisfaction of taking the first four places.

Arriving back in the USA after the UK versus USA match races, Kel decided to fix

The Beast

the exhaust problem that was still endemic. New pipes from the factory had not helped at Imola; a fundamental re-design was needed. In order to provide the extra room necessary for the pipes under the engine, the left-hand pipe was routed up over the crankcase behind the carbs to exit just below the seat on the right-hand side of the bike. This permitted conical pipes to be fitted, eliminating the pipe fractures. The rest of the road-race season in the USA was totally dominated by Kel's team of Kenny Roberts, Gene Romero and Don Castro. The only blot in the perfect copybook was Gary Nixon's win (ably assisted by 'mechanic' Erv Kanemoto) on a TR750 Suzuki at Loudon in a rain-soaked, totally disorganized meeting.

The same story was being repeated throughout the world with only occasional defeats at the hands of TR750 Suzukis and, even more infrequently, Kawasaki KR750 wins. The first model was not without component failures. The sleeved cylinder head nuts would sometimes fracture causing loss of coolant and occasional seizures. There were a few main bearing seizures not long after the new bike was started up, caused by incorrect crankshaft assembly in the factory. The driven gears of the oil and water-pumps broke occasionally, due to material defects. A number of private owners complained of overheating problems, putting the cause down to the thermostat control, whereas it was probably due to the partial blocking of the radiator by the standard Yamaha fairing. And of course there were those flat-sided exhausts, being increasingly rejected in favour of clones of the Carruthers pipes.

Imola 200 mile race in April 1974. Spot the riders without a TZ750. Sheene (47) on a Suzuki and a couple of Kawasakis, but Agostini (10) pipped Roberts (1) for the win.

The Beast

Notwithstanding the almost total superiority of the TZ750, later to become designated the TZ750A despite its 700cc, Yamaha decided to go for the full displacement by increasing the bore to 66.4mm. In total, two-hundred and sixty 700cc TZ750s were produced, two-hundred and thirteen of them masquerading as TZ750As and forty-six as TZ750Bs. In fact there was no difference between these machines, except the manufacture date of the Bs, which was October 1974. After the initial forty-six 750Bs had been produced, the production run continued to produce a total of sixty-five full displacement Bs. The change to the full 750 brought the obvious change of cylinders and pistons and little else. The water-pump, a couple of gears (there were two recorded cases of gearbox seizure during 1974) and the rear chain tensioner were beefed up. An attempt was made to reduce the exhaust fractures by reinforcing the most vulnerable parts of the pipes with extra gusseting. This bike remained in production in September 1975 when a further forty were produced and sold as TZ750Cs.

Although Yamaha did not seem to be developing the 750 that now had complete domination of 750cc racing in Europe and the USA, this was not strictly true. The increased engine displacement was good for another 10 to 15bhp bringing the power of the works bikes up to around 130bhp, although Yamaha wouldn't admit it officially. Towards the end of 1974 a 750 in the GP 500 monoshock frame was spotted undergoing tests in Japan, and sure enough there were three monoshocks available used at Daytona in March 1975. The lucky recipients were Agostini, Roberts and Canadian Steve Baker, representing the three Yamaha teams competing from Europe, USA and Canada respectively. Steve Baker had some initial problems setting up the damping on the De Carbon shock resulting in rear wheel hop during practice. Kenny Roberts was pleased with the new bike from the start and underscored this by setting the fastest

It took a year before the monoshock chassis for the GP 500 was available for the 750. At Daytona 1975, Agostini received his bike, also featuring the up-and-over routing of the left-hand exhaust pipe.

The Beast

qualifying time. In particular, the revised single-shock rear suspension seemed to have significantly reduced the high-speed wobble suffered by the first TZ700s. At the start of the race he scorched away from the field only to suffer clutch failure after a few laps, leading to another 'did not finish'. It was to be a couple more years before Kenny Roberts could throw his Daytona jinx. Romero won the 200 miler on a twin-shock Carruthers bike, followed home by Steve Baker on the first monoshock. Agostini experienced carburation problems and finished fourth. TZ750s took the first sixteen places; it was total domination. Third place went to one J. Cecotto riding a standard twin-shock flat-piped TZ.

Johnny was passing through on his way to Europe and a shattering 350 World Championship in his debut year of GP racing. He kicked off his European campaign with a win at Imola, after Daytona the second round of the official FIM F750 series. Cecotto won both of the 100-mile legs quite comfortably after Agostini suffered broken cylinder head studs in the first leg and Kenny Roberts called it a day, still suffering from a high-speed crash during the Trans-Atlantic Trophy races at Oulton Park a few days before. Later in the year Cecotto was

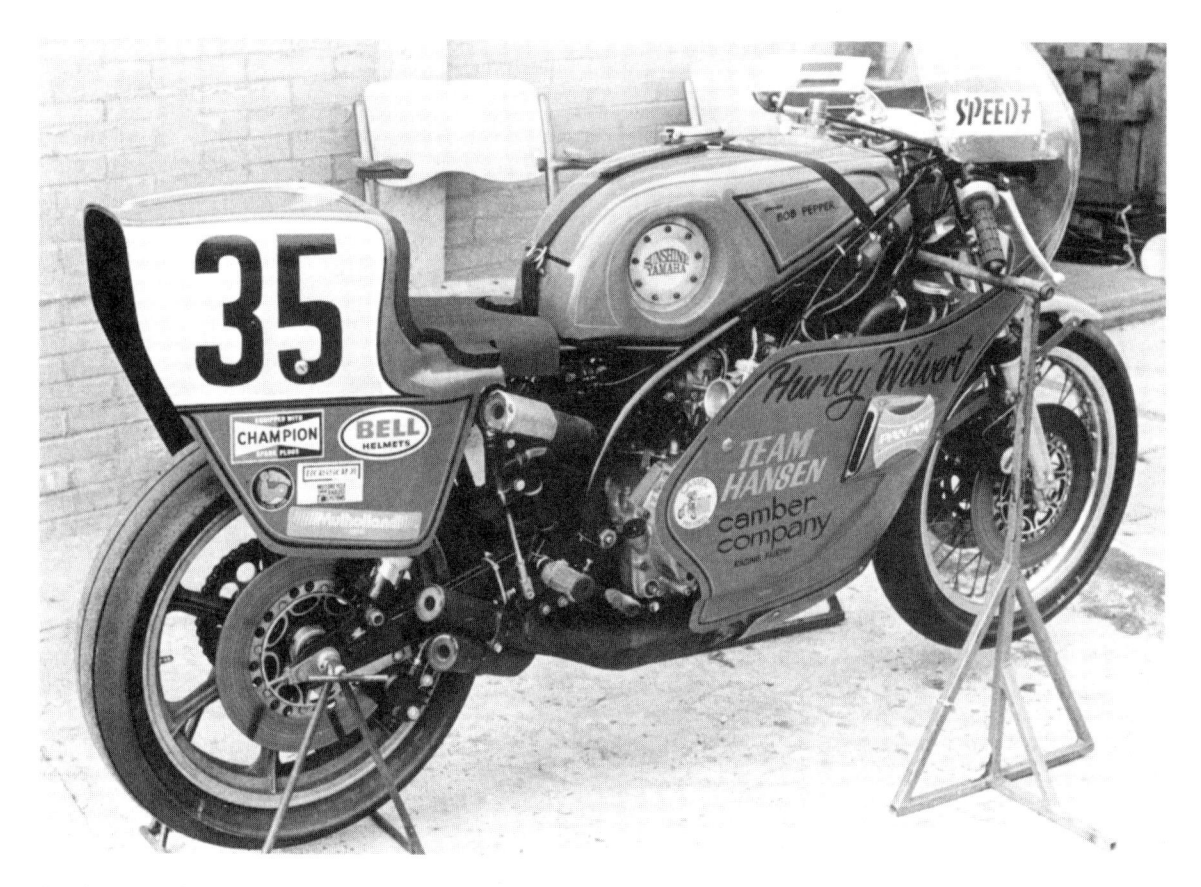

Professional bike preparation was not limited to the factory teams. Privateer Hurley Wilvert competed at Daytona 1975 on a TZ750 looking every bit as capable of winning as the factory machines.

also provided with works monoshock machines and works mechanic Vince French, but his attention had passed to the 350 crown and the results were disappointing.

TZ750 MILER

Meanwhile back in the USA, Yamaha were looking at how best to retain the AMA number 1 plate Kenny had won in 1973 and held for 1974. The championship consisted of twenty-one races, but only four of these were road races. The rest were flat track milers and TTs, on which the Harley-Davidsons excelled. When the Atlanta road race was cancelled in June, things looked bleak for Yamaha. Their 650 twin-milers were barely competitive, and attempts to match the performance of the Harleys had led to it becoming unreliable. Frustrated by the looming loss of his number 1 title, Kenny shipped all of the Yamaha flat-trackers back to Los Angeles and ordered Kel to build him 'a rocketship'. Kel took him at his word and within ten days had built the most monstrous machine ever to be seen on a dirt track, the TZ750 Miler.

Kenny was not actually the first person to be confronted with the challenge of riding this bike. The beast was conceived one winter afternoon in January 1974, when Steve Baker and Bob Work stopped at a friend's house in California after picking up Baker's TZ 700 from the Anaheim Dealer Show. The friend was Doug Schwerma, renowned for his frame design for flat track machines. Within minutes of seeing the TZ, he was measuring it up to determine if it could be shoehorned into the frame he built for the four-stroke Yamaha 650/750 twins. Crazy as it had seemed at first, the closer they looked the more feasible it became. Schwerma contacted Yamaha and asked to borrow an engine to design a frame around. Pete Schick, the racing manager, thought the idea was crazy and would not co-operate. The 650/750 bikes were still competitive at that time.

Schwerma was convinced that it was possible and that a rideable bike would be produced. He'd already built frames for Erv Kanemoto, who had slotted Kawasaki 750 triple engines into them and won the non-national Stockton mile with the resulting bike in July 1975. Eventually he got his hands on a TZ engine in the spring of 1975 and built a bike that was ridden by Rick Hocking to third place at the Ascot Park TT in July. Yamaha were impressed and placed an order for five more frames, for immediate delivery, in time for the the Indianapolis Mile on 23 August 1975. Skip Aksland, Steve Baker, Kenny Roberts and Randy Cleek were entered to race the machines. Kenny's bike had the same engine he had used to win Laguna Seca three weeks earlier, using the old flat-sided pipes to provide more ground clearance. The one concession to the outrageous power delivery was a kill button on one cylinder to chop 30bhp and give the rear tyre some chance of hooking up.

Practice was not a bed of roses. Steve Baker was having trouble getting his old dirtbike style of riding dialled in. The engines of both Cleek and Hocking seized. Skip Aksland and Kenny Roberts were having tyre trouble... This is how Kenny recounted it later to Cook Neilson.

> I never saw the TZ Miler until it turned up at the race track. There were four of 'em. I thought it would be really funny if all four of us fired 'em up, went into the first corner, found out at the same time that we couldn't turn 'em, and all four of us smacked into the wall.
>
> I was the first one on the track with one, and planned to take it easy, since it was a brand new bike. Didn't take it easy too long. It was geared for 145mph at 10,500rpm in sixth gear. In the third turn I really gassed

The Beast

Kenny Roberts in the tech line, about to blow the AMA inspector's mind with the most powerful miler in the history of the sport.

it, and the thing jumped up on the back wheel. Almost freaked everybody out. It was *fast* but it wobbled all the way down the straight, because it was lifting the front wheel all the way down the straight. It had a good power band, but it spun the tyre *bad*.

In the heat race I got something like fourth. That was when I decided to shift it. I took two teeth off the back sprocket, which would have been 150mph in top gear, and it was pulling between 10,500 and 11,000 in the Main – it was going 150mph at the end of the straight!

To get it stopped, I had to throw it down on its side with the brake on, and bounce it off the cushion. There was a berm in turns Three and Four, and I could go through there faster than the Harleys could. But in One and Two there wasn't such a big berm, and I had to run out against the fence, in

The Beast

the muck. I would lose as much as a half a second a lap, depending on the drive in One and Two.

Towards the end of the race I was in third, and I didn't know how I was going to get a drive strong enough to win. I was 30 yards behind the leaders Corky Keener and Jay Springsteen. They were swapping the lead, each wanting to be second coming out of the last corner. Corky had no idea I was behind 'em, but Springer did. About five laps before the end of the race he looked back and saw me. I thought, damn, he *saw* me. And I didn't want 'em to see that I was hiding out back there, planning to do a last lap number on 'em. When Springer passed Corky, he signalled that Number One – me – was right behind. But Corky didn't understand. On the last lap they were still playin' around. Corky wanted to be second. Coming

Corky Keener and Jay Springsteen eat the dirt thrown up by the most unlikely miler in the history of the sport, the TZ750 geared for 150mph on the straight.

out the last corner Springer was in front, and Corky used the draft to pull up alongside. So I had a two-bike draft. I held the throttle really steady, got a good drive, and when I pulled the clutch in to catch the next gear, the tyre stopped spinning, got a really good hold, and just went by 'em. I won by about a foot. That was the only time in America that I've seen a crowd go nuts.

There's no doubt in Roberts' mind that this was his greatest race, a triumph of courage, skill and strength mixed with a touch of luck. It was to be the only time a TZ Miler won a national flat track race. There were two other outings, the first being at the Syracuse Mile on 7 September, when a slippery track eliminated all the Yamahas which gave Harley a clean sweep. Kel spent some time trying to improve the bike before the last Mile race of the year at San José on 21 September. In order to get more weight on the front wheel, he moved the engine forward and lower in the chassis. This was no small task, involving reversing the cylinder head so that the water-pump and plumbing didn't foul the frame tubes. It was to be in vain. The first practice lap involved an almost continuous enormous tank-slapper, that was sorted out enough for Roberts to win his heat and qualify sixth fastest. Gambling that a slick tyre would be better than the treaded tyres he had been running, it proved a mistake and he pulled in stating that it was unridable. With that retirement, he lost the AMA number 1 plate to Gary Scott, although Yamaha pipped Harley to the manufacturers' title.

Although not apparent at the end of the 1975 season, this was the end of the TZ Miler. A throwaway comment after the Indy race that 'They don't pay me enough to ride that thing', prompted the AMA's professional rules committee to rush through legislation that winter limiting the bikes to two cylinders. The bikes were stripped, the engines going back to winning road races, while the chassis were scrapped. As Kenny Roberts said later 'I had no problem with riding that Miler, and I never felt that it was dangerous for myself. I just couldn't see fourteen other people on them.'

In the USA, the TZ750 was invincible in 750 and 1,000cc road racing. In the rest of the world, there were still isolated pockets of resistance from factory supported Suzukis and Kawasakis. Barry Sheene won three rounds of the FIM F750 championship, which was almost enough for him to pip ultimate winner Jack Findlay on a TZ750. In the UK, most 750 races were between Barry Sheene and Mick Grant and Barry Ditchburn, who rode factory water-cooled KR750s. The 1,000cc Isle of Man TT race was won by a TZ, but a 350 not a 750, which was a little bit overpowered and too heavy for the street circuit. This did not worry Yamaha; 750 racing in all countries was dominated by their machine. They had something up their sleeve to finish the other two-strokes once and for all.

TZ750C AND OW31

In September 1975, the factory cobbled together another forty machines for sale to the public as TZ750Cs, identical to the full-displacement TZ750Bs. They also announced their withdrawal from GP racing as a factory team, having won the 500cc crown for Ago. Although the 750s for public consumption did not benefit from radical improvements, the factory had already prepared five bikes for the works riders for 1976. These were passed out to Johnny Cecotto, Giacomo Agostini, Steve Baker, Kenny Roberts and Hideo Kanaya and became renowned under their factory designation of OW31.

Essentially, this was a breathed-on 750 block housed in an ultra-light 500cc GP monoshock frame. A lot of titanium and magnesium was evident resulting in a weight reduction of around 40lb (18kg). The steering head stem was made of a tough aluminium alloy, the lower triple clamps were magnesium, the brake calipers were aluminium with titanium pistons and even the front fairing stay was titanium. The monoshock was 4.5lb (2kg) lighter than the previous year. The engines were ported to Kel's magic Daytona measurements. Kenny Roberts was timed at 182mph (293km/h) during practice, but surprisingly he was not totally happy:

> The first time I saw mine I wanted to put it back in the crate and send it back to where it came from. Sure it was a nice bike, but we didn't need it. All it was going to do was make people mad.

At the end of practice, the four entered OWs (Ago had stayed at home) were the four fastest qualifiers. Only one and two halves finished. Steve Baker's ignition failed quite early. Hideo Kanaya's rear tyre was shot after about two thirds of the race and he pulled in for a new one to eventually finish seventh. Kenny's tyres were going the same way and on the forty-fifth of fifty-two laps his rear tyre punctured, causing him to use some straw bales as a berm to keep on the track and return to the pits. After replacing

Agostini's OW31 was not put to good use in Europe in 1976. With a reported 140bhp and 330lb (150kg) the OW only lived up to its promise in the USA in the hands of Steve Baker and Kenny Roberts.

The Beast

Everyone loves a winner. Cecotto has just won the 1976 Daytona 200 miler. However it was downhill for the rest of the season.

the rear tyre he finished ninth. Johnny Cecotto made it a second win in a row. Even his race was not trouble free as the centre expansion pipe persistently grounded and eventually developed a hole. Vince French had foreseen this potential problem and had jetted that cylinder slightly richer to ensure a finish. Unfortunately for Johnny, this turned out to be the only good result of a very disappointing season in both the F750 and other Championship classes.

It was not to be a spectacularly successful year for the OW31s. They won the other three US road races contributing to the AMA National Championship. Two wins went to Steve Baker and one to Kenny Roberts, who slipped to third place in the final championship positions. Baker also won the F750 races at San Carlos in Venezuela and Imola, but fortunately for Yamaha, the San Carlos result was deleted from the official championship after a dispute about the timekeeping. If the results had stayed, Gary Nixon and Erv Kanemoto, using works Kawasakis they had bought from the factory, would have taken the championship by a single point. As it was, privateer Victor Palomo, riding a number of different production TZ750s, took the title, humbling the Yamaha and Kawasaki works bike owners in the process. In truth only half his machine was Yamaha, as he had chosen, like many other top riders, to use a Nico Bakker cantilever frame instead of the standard Yamaha chassis.

The Beast

Kanaya's OW31 ate its rear tyre during the 1976 Daytona race. Its replacement dropped him to seventh place.

TZ750D

Yamaha were now convinced that they would need to substantially update the TZ for 1977. They decided to sell the TZ750D as a replica of the OW31, but without the exotic metals the factory machines had used. Apart from some changes to the crankshaft and pistons most of the engine was unchanged from the C. New exhausts included integrated silencers to comply with the tighter FIM noise regulations that had killed the MVs off once and for all. The chassis looked very much like the OW31s of 1976, but it was only 4.5lb (2kg) lighter than the old twin shock frame. The same general cradle design was maintained, but an extra tube ran under the front of the engine between the cradle loops. Some of the bracing tubes were re-positioned slightly, to make room for the monoshock suspension. At 12.5cm, the new suspension provided almost double the amount of rear-wheel travel offered by its twin-shocked predecessors.

In total only thirty TZ750Ds were offered for sale for 1977, costing $5,195 in the USA and £6,500 in the UK where five machines were bought by the top UK privateers. Perhaps the exotica of the OW31 had raised their expectations as they were initially disappointed by the limited improvement with respect to the C model. In particular, quality control had been very poor, with machines shipped with partial seizure due to piston/cylinder mismatches. Footpegs and steering dampers broke very early in the season. There was some difficulty setting up the monoshock suspension correctly. Most disappointed of all were the British sidecar crews that had just bought the engine, a snip at £3,200, but virtually identical to the 750C unit.

As always, Daytona was the start of the new season and the factory teams were working with essentially the same machines as the previous year, factory special OW31s. Detailed changes included Powerjet Mikunis on Cecotto and Baker's machines, while Kel and Kenny stuck to the Lectrons they had started using in 1976. There were four machines available to Baker, Cecotto, Roberts and Australian Warren Willing. Cecotto's hopes for a hat-trick were dashed when he retired on lap four with transmission oil leaking on to the track. By the end of the first leg, Steve Baker had pulled out a 28-second lead over a disappointed Kenny Roberts who seemed to be slightly down on power compared to Baker. Shortly after the end of the 100 miles, it started to rain and didn't let up, causing the second leg to be cancelled, and Roberts had once again started Daytona as favourite and been unable to make it stick.

Daytona was interesting for the diversity of specials that had been built around the 750 engine. A particularly interesting machine was Erv 'If you can't beat 'em, join 'em' Kanemoto's TZ750 with belly-pan fairing to the rear wheel axle and hand-beaten aluminium tail fairing. It was built for Gary Nixon, who never got the opportunity to ride it after breaking his wrist during practice when his TZ250 seized. This and a number of other machines used C and J monoshock frames that were built to the same basic Yamaha geometry, but were lighter than the standard D chassis. The engine underwent some attention during 1977 as Erv realized that it was over-scavenging as the transfers directed the charge up towards the spark plugs, and away to the gaping exhaust port. A combination of shorter exhaust pipe headers to reduce the volume of the pipe and grinding flatter roofs to the transfer ports, to direct the charge from both cylinder sides towards each other, alleviated the problem somewhat.

FINAL GLORIES

After Daytona, Steve Baker went to Europe for the 500cc GPs and the F750. He was to be spectacularly successful in the 750 series, winning the championship with ease in a near-perfect performance, resulting in five wins, three second places and two third places in the eleven-race series. This was Yamaha land if ever there was such a place. Of the forty-one riders getting points during F750 races, only two were not Yamaha-mounted, namely Greg Hansford and Yvon DuHamel on the veteran KR750s. They did especially well at Mosport in Canada finishing first and second on a slippery treacherous wet circuit.

Although no-one realized it at the time, the TZ750D marked the end of the development of the big TZ. Yamaha built another 162 OW-31 replicas, identified as TZ750Es for 1978 and TZ750Fs for 1979. The FIM declared the suspension of the series from 1980 at the same time scrapping the ruling that the bikes must be homologated for 200 units. Ironically, this let Suzuki campaign bored-out RG500 GP machines, and could have led to more interest in the class. But it was all too late and Yamaha closed down the production line after a total of 567 units had been produced. The last TZ750F machine to be sold went over the counter in January 1983.

The TZ was no longer being developed by Yamaha, but it continued to win races. Kenny Roberts finally won Daytona in 1978,

The Beast

Steve Baker was practically invincible on the 1977 YZR750.

lapping every rider in the field at least once. This was the year of the infamous AMA restrictor regulations that had been introduced in an honest, but misguided attempt to limit the power being produced by the 750 four-cylinder engines, to even them up with the triples. By enlisting Yamaha's help in specifying the restrictor's dimensions, the AMA gave the factory a seven month headstart in designing the most effective inlet tract. The advantage was clear as soon as practice started, with Kenny Roberts easily fastest, whilst experienced tuners like Erv Kanemoto and Don Vesco experienced persistent seizures. Kenny actually preferred the bike with the restrictors and ran it during 1978 in many of his F750 races in Europe. The restrictors were finally dropped in 1982, by which time Honda were campaigning 'killer' four-cylinder 1,000cc four-strokes. It seemed like a good idea at the time ... Johnny Cecotto finally won the F750 crown in 1978, transferring soon after to four-wheel racing. The year after was to be Frenchman Patrick Pons' year with wins at Daytona and the last F750 Championship title, the first world motorcycle title won by a Frenchman.

As late as 1983, the TZ was still winning races in the US against both large four-strokes from Kawasaki and Suzuki and the new three-cylinder factory 500cc GP Hondas. Miles Baldwin came very close to wining the US road race title that year on a ratty TZ750E tuned by the *doyen* of technical journalists, Kevin Cameron. It looked so battered, that AMA officials gave him a lecture on hurting the sport's image.

The TZ750's last moment of glory came midway through that year at the fifth round of the American Formula 1 bike series. On Sunday 26 June, at the Pocono International Raceway in Pennsylvania, Gregg Smrz headed home Doug Brauneck and Miles Baldwin to give the Yamaha TZ750 a 1–2–3 in the race and a 1–2–3 in the championship points with three races to go. These were four-year old machines ridden by gifted

The Beast

It might have looked like a rat bike but Miles Baldwin took his 1979 TZ750F to the runner-up position in the 1983 US Formula One championship.

privateers, armed solely with the accumulated knowledge gained from the tens of thousands of racing miles consumed by the TZ. On that warm afternoon, almost ten years to the day that Kel Carruthers had been pushed out on the track for his first hesitant lap on the 'Beast', the TZ750 was still the king of road racing motorcycles.

8 The Tenacious Twins
TZ250 1981–93

As the 1980s opened with Roberts' third successive 500cc crown, his was one of the few 'big-budget' teams to be found in the paddocks at European GPs. Since the inception of the GP series in 1949, only factory teams had large amounts of money available to support their world championship challenges. Until the 1970s only very limited sponsorship from oil companies and others connected with the world of motorcycles had been available. Privateers bankrupted themselves to participate in the 'Continental Circus' for a few years, only to move into one of the wealthy works teams if they caught the eye of the team manager, or to call it a day when the funds ran dry. Prize money levels defined by the FIM were woefully inadequate for the mid-field riders, the ones who needed it the most.

Since the riders had such tight budgets, Yamaha were forced to keep the prices of their production racers at a level that these riders could afford to pay. As long as there was a street model on which to base the racer, this could be achieved and so it remained until the mid-1970s. It took Suzuki to break the mould with their RG500 that was offered to a limited number of riders in 1976 for the princely sum of £4,500. This was the first full-blooded production racer with no street machine pedigree, since the last Manx Norton had been rolled off the Bracebridge Street assembly line in 1961. Its arrival coincided with the appearance of the first stickers of the non-industry big buck sponsors that were to change the face of the sport during the 1980s. Marlboro stickers appeared on Agostini's bikes during 1975 and Gaulois started sponsoring the French Sonauto team around the same time. These sponsors brought with them the Francs, Lire and Pounds needed to purchase the increasingly complex 500 and 250 racers that were filling the starting grids. The 1980 TZ500G cost £12,000, about the same as the RG500 of the time, with the TZ250G costing £3,900. Four years before, a TZ250C, the standard mount of the GP riders cost just £1,500. The 1981 TZ250H was priced at £4,500. Racing was becoming big business.

TZ250H

Riders were more willing to pay good money for the 250H than they might otherwise have been, as it was a totally re-designed motorcycle. The technology that had been tested and de-bugged on the 500 factory machines now found its way into the hands of the world's privateers. The powervalve had arrived in the public domain. All the reasons why it had been so desirable on the 500 were equally valid for the 250. The 250G had been fast but flawed, requiring new pistons after every race, and plumbing the depth of its owner's pocket. The total re-design of the engine that accompanied the application of the powervalve, enabled the factory to provide a package with a better

The Tenacious Twins

The TZ250H of 1981 had undergone a total engine re-design, the first of the TZ250 series.

balance between speed and reliability.

Following the lead of the 500, the cylinder castings were separate items, base-mounted to the crankcase, with a bore and stroke of 56 × 50mm, marking a return to the dimensions of the legendary TD2B. Still wishing to maintain wide open spaces on the inlet side of the engine, Yamaha's fix for piston reliability was to design the engine to run backwards. Now the power strokes of the engine drove the skirt of the piston up against the unbroken wall of the cylinder underneath the exhaust port, providing support and longevity, and enabling the inlet port to be opened a few more millimetres. A jackshaft was placed behind and under the crankshaft to reverse the direction of the engine before it reached the transmission.

On the old engine, the jackshaft would have been almost submerged in the oil sloshing around the sump to provide gearbox lubrication. With the shaft turning at engine speed, this would have resulted in an unacceptable loss of power. To avoid this, the sump volume shrank to 500cc and a pump passed the oil to lubrication points in the transmission.

There was no longer a single crankshaft. The left-hand crankshaft had a gear pressed on its inboard end, which drove the jackshaft behind. The right-hand crankshaft had a smaller diameter gear pressed to its inboard end which provided a spline fit with teeth cut on the inner surface of the gear on the other crankshaft. The power tap from the centre of the engine, coupled to the lowered piston speed of the short-stroke engine, tougher crankpin and lighter crankshaft flywheels resulted in a significant reduction in the bottom-end problems G owners had experienced.

The ignition maintained its traditional position on the outboard end of the left-hand crankshaft. The right-hand crankshaft drove the mechanical governor for the powervalve mechanism. Yamaha had decided not to provide the privateer with the electrically driven mechanism found on the TZ500. Instead, they designed an ingenious system of ball bearings, curved cups and springs that converted the movement of the ball bearings caused by centrifugal pressure of the rotating crankshaft into vertical movement of the powervalve linkage. This in turn caused the two valves, linked via an Oldham coupler to rotate in their housing within the cylinder. Up to 8,750rpm, the valves remained fully closed, projecting 6mm into the exhaust port. By 11,000rpm, the port was fully open with the engine blowing the spent gas through the gaping exhaust port for all it was worth.

It wasn't only the exhaust side of the engine that came in for improvement. A new type of Mikuni carburettor was fitted to the H. Since 1976, a small company in Michigan had mounted a successful challenge to Mikuni's hegemony in the racing world. Norm Quantz, chief designer at Lectron, had produced a new flat-slide carburettor that had a single needle that controlled fuel flow throughout the range of engine speeds. Most other carburettors, including Mikuni, had separate idle and main metering systems and the transition between the two as the throttle was opened was always imperfect. The main culprit was the throttle slide cutaway, whose purpose was to help lean out what would otherwise be too rich a mixture at idle speeds. Unfortunately it also reduced the pressure difference over the main jet as the throttle opened, causing the fuel to be only partially atomized. The Lectron carburettor had a guillotine slide with almost no cutaway. The single metering rod had a flat taper ground on the cylinder side which metered the flow throughout the range of throttle positions. Users of Lectrons were rewarded with quicker, cleaner, throttle response, and the word was soon out that the Lectrons represented 'bolt-on horsepower'. By 1981, Mikuni had adopted the motto 'If you can't beat 'em join 'em' and introduced the 'ZC' with no idle system and a cylinder side cutaway of the cylindrical slide, exposing the main jet to crankcase vacuum. This, combined with their Powerjet system, gave them the edge over the Lectron which disappeared into the mists of time within a couple of years.

After such an exhaustive re-design of the engine, it was a surprise that there was enough money over to do some chassis work as well. The frame was 6.5lb (3kg) lighter, the engine positioned slightly further forward and the rake reduced to 24.5 degrees. The rear aluminium shock absorber became more tunable, with bump damping adjustment added to the rebound adjustment previously possible. Just to keep everybody's feet firmly on the ground, the machine was offered with wire wheels which had not been seen on a bike actually racing since the middle 1970s. Maybe the money did run out after all ...

The 250H was expensive but it was worth the extra money. Most of the Yamaha GP contenders rode it in 1981, the bike proving to be only slightly slower than the dominant works Kawasakis of the time. Carlos Lavado had a good start to the season, taking several rostrum places, until he crashed in Italy and broke his leg. Several riders of 250H machines were given engines with electrically controlled powervalves, including both the Venemotos and Sonauto teams. Lavado, Sarron and Espié were noticeably more successful than other Yamaha riders as a result.

The Tenacious Twins

Martin Wimmer and Carlos Lavado were to be the core of the Yamaha team for much of the 1980s. Here Wimmer is congratulated by Anton Mang after winning his first GP, the 1982 British GP at Silverstone.

Yamaha were going through hard times at the start of the 1980s. The world recession of 1981 hit them quite badly, although it took them until 1983 to discover how badly their sales had been hit. The poor state of the street bike business was reflected in their lack of success on the race track. Although the 250H had been an improvement over Yamaha's previous parallel twin, the arrival of the Rotax tandem twin engine in a variety of different frames, cast a shadow over the TZ. The first Rotax twins had been raced in the UK during 1980, in the form of Cottons ridden to six wins out of eight races in the Vladivar Vodka series of international 250 race meetings. Steve Tonkin, Clive Horton and Steve Cull ended up taking the first three places for Cotton in the final championship results. By 1981, a number of Rotax- engined machines started appearing on the GP grids, including the British Armstrong and Spondon, Swiss Egli and Spanish Sirokos. They were not yet competitive at GP level, but once again cleaned up at British international meetings.

TZ250J

Things were going horribly wrong for Yamaha in the 500 class as they struggled to find the right replacement for their obsolete piston-ported 500. All the race shop's energies were directed to building the new

The Tenacious Twins

bikes that were under development for the 500 class battle with Suzuki and Honda, who had switched to two-stroke power for their attempt on the 500 crown. Consequently almost no changes were made between the 1981 TZ250H and the 1982 TZ250J. The exhaust muffler exit pipe was curved downwards, some alternative gearbox ratios were made available and the front forks underwent some minor modifications. Many of the most competitive 250 Yamaha riders were complaining of the amount of front-wheel patter they were experiencing, this more than anything limiting their cornering speeds. Some were making the move away from the traditional 18in to 16in front wheels, making the bikes easier to turn through fast corners, but with inherently more nervous handling. It was hoped that the reduced weight of the front wheels and correspondingly smaller disc brake rotors would also reduce patter. It was to remain the racers' scourge until advances in front fork technology by White Power, Öhlins, Kyaba and Showa, all but eliminated it during the second half of the decade.

Despite the lack of development of the twin, 1982 was to mark the return of the 250 riders crown to the Yamaha fold after an absence of eight years. Surprisingly it wasn't the semi-factory teams of Venemotos and Mitsui Germany, running electrically driven powervalves, flat-side magnesium Mikunis and re-designed exhausts that achieved it. Jean-Louis Tournadre amazingly stormed through the season, with consistent rostrum positions, and took the title from Anton Mang in only his third year of GP competition. It was an excellent performance, although his single victory of the season at the French GP that was boycotted by the majority of the 250 riders, was the decisive factor in his world championship triumph. Sadly, his aspirations to defend the title in 1983 turned to dust as he failed to gain a single point throughout the whole year and thoroughly depressed by the whole thing he turned his back on the motorcycle sport at the end of the year.

TZ250K

Much of the stuff German Martin Wimmer and Lavado had been running in 1982 was being considered for inclusion in the standard 1983 TZ250K. Not everything made it, the most visible omission being the electric powervalve, which was still only offered as a very expensive kit. However the good news was the new cylinder heads, 38mm roundslide Mikuni carburetors, magnesium clutch cover, compacter clutch with the removal of a single plate, new exhausts, lighter frame, new springs on both front and rear suspension and a narrower fairing. The whole setup was good for a drop in weight of 6.5lb (3kg) and the addition of a couple of bhp. The improvements were not enough for vast numbers of H and J owners to make the change, especially as the price took another hike to £5,250.

For 1983, Yamaha's season went according to plan, at least for the 250 class. Carlos Lavado took a comfortable lead in the series at the start of the season and ended twenty-seven points ahead of the Sonauto Yamaha of Christian Sarron. Kawasaki's days of dominance had come to an end with the factory's withdrawal at the end of 1982, and the Rotax challenge had still to materialize. Manfred Herweh's win in Austria on his Real Rotax, with third places in Italy and Yugoslavia, suggested that Yamaha would have a fight on their hands in 1984.

TZ250L

Martin Wimmer was the first GP rider to ride the 1984 TZ250L. At the 1983 Dutch TT, he had crashed the bike in practice and

The Tenacious Twins

used his kitted K model to claim fifth place. A week later at Spa in Belgium, Martin took it to sixth place despite suffering from a badly swollen elbow, the legacy of his Assen crash. As far as Wimmer was concerned, the L he was riding was simply a new chassis. The engine was moved slightly further forward, and the castor steepened to place more weight on the front wheel. This change alone would have caused the bike to be a very nervous steerer, tank slapping at the slightest opportunity. Some straight line stability was achieved by increasing the wheelbase, by lengthening the swinging arm by 40mm. The front fork was again reworked and 2mm added to the stanchion diameter. Again the patter was reduced, but not eliminated entirely. The rear swing arm was strengthened by extending the box gusseting between the upper and lower arm a few centimetres along the length of the tubes toward the rear axle.

These pre-production units formed the basics for the production version of the 250L that became available in February 1984. The engine now contained the 250K kit parts as standard, featuring the electrically driven powervalve, new exhaust, re-jetted carburettors and pistons with a shorter skirt on the inlet side. Crankpin and con-rod were

Christian Sarron's Sonauto 250L dethroned Lavado's Venemotos 250L in 1984. The Hummel cylinders on Sarron's machine gave him a speed edge that Lavado's razor sharp cornering couldn't compensate. He crashed four times while leading a GP.

The Tenacious Twins

also strengthened to pre-empt any crankshaft problems with the engine that now produced over 60bhp. A new piston design was introduced featuring a smooth internal surface underneath the crown. Tests had shown that the ribbings, rather than the strengthening of the piston, were sources of metal stress, leading to the formation of cracks. Once again the new machine was introduced with a massive hike in price to £7,500, meaning there were few takers at any level of the sport below the GP contenders. Gone were the days when Yamaha could afford to subsidize the cost of racing under the creed of 'racing improves the breed'. With the relationship to the street machines long gone in all but spirit, every minute of development time, every gramme of metal had to be payed for by the customer.

To make matters worse, these expensive TZs were not even very good. Several teams, including Mitsui Germany and the new Roberts Marlboro 250 team, with rookie Wayne Rainey getting his first taste of GP life, were given special kits by Yamaha with guillotine powervalves, wider exhaust ports and cylinder heads with almost no squish band. The exhaust porting now consisted of a main exhaust and two auxiliary ports either side. These kits proved to be a total disaster on their first appearance at the Italian GP at Misano, causing detonation and holed pistons. Wimmer rejected the kit in preference for cylinders made by ex-50cc rider Hans Hummel, whose after-market TZ cylinders had been building an enthusiastic following in the paddock since the previous season. By mid-season, almost all of the

Lavado's 1984 machine was externally a standard 250L. His third place in the final standings is a testament to his enormous riding skills.

127

The Tenacious Twins

Yamaha riders were using Hummel cylinders. Interestingly enough, they were not fitted with powervalves. Hummel had a special relationship with the Sonauto team, which also extended to sidecar drivers Alain Michel and Egbert Streuer. He provided them with his most recent design for evaluation, hence Christian Sarron was riding with Hummel cylinders with a guillotine powervalve from the Yamaha factory kit. The guillotine only functioned on the central main exhaust port, the auxiliaries remaining open at all engine speeds.

The Roberts team had also tied up with Austrian tuner Harald Bartol, who had been working on 250 cylinders with reed valves. Honda's success with the three-cylinder RS500 had sparked a resurgence in interest in the reed valve, but Bartol's design was not actually raced during the season. Further experimentation by the team led to them adopting a twin-spar aluminium frame for the late season British GP. The design was based on that used by the Yamaha YZR500, at last offering the advantage of pukka rising rate suspension instead of relying on the varying pitch of the wound spring on the rear damper. Twin spar frames

The shape of things to come. Rainey's spar frame was a clone of the OW76 500, essentially the same design appearing two years later on the 250s. Check the 'HH' on the Hummel Cylinders without powervalve.

were beginning to appear on a number of machines including the Italian Suzuki 500 team of Roberto Gallina and the Armstrong team running 250 Rotax engines. They represented the first shots in a chassis revolution that was about to engulf the world of racing motorcycles.

Christian Sarron won the 1984 250 world title for Yamaha riding a TZ250L with Hummel cylinders. It was the last time a TZ would win a world championship title during the 1980s and perhaps ever. It was a season of very close racing between Lavado, Sarron, Wimmer, Anton Mang and Jacques Cornu on Yamahas and Manfred Herweh and Alfonso 'Sito' Pons riding Rotax based machines. Falls by Herweh at Silverstone and Salzbürgring denied him the title. The machines had been evenly matched and had provided some of the most entertaining racing the class had been able to offer for years, but all this was about to change. Joining the battle to de-throne Yamaha from their domination of the class came Honda who, for the 1984 season, had provided a few riders with production racers based very closely on their street 250 sold in Japan. These first RS250s were painfully slow and heavy, but at the season's end, Honda announced their intention of producing a batch of red-hot racers with aluminium spar frames and an engine providing 70bhp. Freddie Spencer would be getting a super red-hot version in his attempt at both 250 and 500 crowns in a single season. Where did this leave the TZ?

TZ250N

The TZ250N that appeared in February 1985 featured a radically re-designed engine. Breaking with the tradition that had works supported riders running equipment the season before general release, both production and factory supported bikes appeared with engines using reed-valve induction directly into the crankcase. Reed valve technology had proved itself in the 500 class, giving first Honda and then Yamaha, engines whose power delivery enabled them to get good drive from slow and medium speed corners. The 500s had also managed to balance the wider power band offered by the reed valve engines, with the slight loss of top-end due to the reed restriction in the inlet tract. Painstaking exhaust design had done the trick. Now the 250s could benefit from the knowledge garnered by the companies in the 500 class, and both Honda and Yamaha 250 racers for 1985 were equipped with reed valves. The only difference between the two was that Honda got it right and Yamaha got it wrong.

The four-petal reed used on the 250N was too small. It is strange that the machine ever reached production in this form, since the Yamaha blue-eyed boys, Lavado and Wimmer, were provided with six-petal reeds from the start of the season. The four petals of the standard reed were simply too heavy. They offered too much inertia to react directly to the signals passed by the engine. Consequently, they would not open soon enough and would close too late to be effective. The reed design forced new cylinders and crankcases, the exhaust side of the cylinders benefiting from the guillotine powervalve and auxiliary ports offered to, and frequently rejected by, the factory riders in 1984. Yamaha compounded the deficiencies of the intake side by lowering the exhaust ports by 1.5mm to help the mid-range they were aiming to improve. The cylinder wall that had been filled by the inlet port, was now graced by two transfer ports of 13mm width. Of course new exhaust pipes and revised carburation were all part of the deal. All round the engine felt dull and down on top-end – a 'droner'. Little else was offered to tempt the potential customer, the chassis suspension changing slightly with revised damping and spring rates.

The Tenacious Twins

Waiting for Spencer? Lavado's brightest moment of 1985 was his win at the Spanish GP when Spencer's exhaust split and he finished down the field.

At national, international and GP level, the 250N was totally outclassed by either Rotax-engined machines or RS250s from Honda. The GP team had a very difficult year, with almost total domination of the class being achieved by Freddie Spencer on his very special reed valved RS250, ably assisted by Toni Mang on his kitted RS250 without reed valve. Wimmer won a wet Hockenheim and Lavado won in Spain, when Spencer's exhaust split, and San Marino with Spencer back in USA with the title safely in his pocket. Non-works Yamaha riders managed a few placings mostly using Hummel cylinders on pre-250N engines, and the Yamaha production racer had lost much of its shine.

The one bright spot during the dismal year was Lavado's pole position at the British GP at Silverstone. The pole position itself was not that special, although there had been precious few that year. The margin of more than one second from second place Spencer was unusually large, but most important, it was set on a new V-twin 250. After a crash late in practice, Lavado had to pull in during the rain-soaked race, his damaged ankle causing him too much pain. In Sweden a week later he again set fastest practice time, half a second ahead of Toni Mang, Spencer having decided to miss the 250 race to concentrate on the 500 race at which he could and did win the championship. During the race the V-twin went off song and Lavado could only manage second place to Mang. With the two GPs under its belt the V went back to Japan for final preparation for a full season in 1986. Yamaha were following the tradition that had been with the company since the 1960s of running completely new machines for a few races to sort out any problems. It takes the extremes of a full GP to shake out any bugs undiscovered during testing. Unfortunately it would also have warned Honda of the challenge awaiting them in 1986.

TZ250S

One or two optimistic souls started theorizing that the next TZ, the TZ250S, would be a V-twin, but they didn't take Yamaha's innate caution and conservatism into account. The V-twin had yet to prove itself, so the new TZ underwent a major evolutionary change rather than a complete revolution. The major

The Tenacious Twins

Martin Wimmer rode for Yamaha for several years but never quite made it to the top. This victory at his home GP on a wet Hockenheim track was the high point of his career.

The story of 1985. Freddie Spencer already had a 10 metre lead at Spa's Eau Rouge. Lavado (3) came home second, Niall MacKenzie slipped off at the end of the first lap. Spencer of course won.

Silverstone 1985 and Yamaha's new Honda beater, the YZR250 V-twin makes its debut. Following in the Yamaha tradition of mid-season launches, Lavado used the bike to smash the lap record before smashing it up in a spill during pratice.

131

problem of the tiny four-petal reed valve on the 250N was alleviated by the use of a six-petal reed, as run by Wimmer and Lavado throughout the 1985 season. Little else of major interest changed within the engine, in contrast to the totally new chassis that dismissed the last vestiges of the machine that had begun development ten years before. The new Deltabox twin spar aluminium frame no longer had the monoshock rear suspension geometry that had been Yamaha's trademark for so long.

The monoshock system that had been introduced on the 1976 C models had offered a theoretical advantage over the twin-shock systems of the time in terms of rear suspension stiffness and greater wheel travel. Single-shock systems that were introduced at a later date were directed more to providing a rising rate of resistance to wheel displacement. This combination of soft suspension at the start of the wheel's displacement and harder suspension as the displacement increased would assist in keeping the wheel in contact with the tarmac and transmitting power to the road. The only way Yamaha could provide this with the geometry of the monoshock system was through the use of tapered-wired springs, difficult and expensive to produce. Other manufacturers built the progression into the rear suspension by designing its geometry to change as the wheel moved. Although more complex to design, the manufacturing of the parts was simple. By 1983, the YZ motocross models had a rising rate suspension; three years later it was the turn of the TZ.

The design of the rear suspension was quite simple. The swinging arm looked very similar to the older designs with the triangulated loop of aluminium tube above the main box-section arm and a massive box brace between the two. The box now had a hole in the middle through which the rear damper poked, projecting down under the swing-arm to one end of a short rocker arm.

The other end of the rocker arm pivoted on a short rod fixed to the bottom of the Deltabox frame. Bolted to the middle of the rocker arm were two short struts attached at the other end to the swing-arm. The top of the rear suspension unit was attached to the middle of the sub-frame extending back from the Deltabox to support the seat and rear fairing. This was a mistake that was to be corrected two models later. On the 250S design, the rear sub-frame had to absorb some of the energy of the rear wheel travel and this caused the welds to the Deltabox to fracture.

Wimmer and Lavado were given these frames to use from the 1985 Belgian GP and they liked them. If anybody had bought the TZ250S, they would have liked them too, coming as they did with light-alloy three-spoke wheels for the first time. As it was, Yamaha were now eclipsed by the Honda RS250 at international and national level, although in the UK Nigel Bosworth, Carl Fogarty and Kevin Mitchell chased the Hondas and Armstrong-Rotaxes hard throughout the year. At GP level, TZs were scarce on the track. Stéphane Mertens managed to accumulate fourteen points and take a thirteenth place in the final positions as a result. Only Frenchman Jean-Michel Mattioli and Austrian Sigi Minich also managed to get their TZs into the points. Fortunately the V-twins in the hands of Lavado, Wimmer and Japanese champion Tadahiko Taira, did a lot better.

YZR250

For 1986, the YZR250, as it was officially named, was the only real full factory machine contesting 250 GPs. Honda had provided a number of riders with NSR250s, which were lighter and more powerful than the standard RS250s, but not as trick as the titanium and magnesium special with which

The Tenacius Twins

1986 was Lavado's year on the V-twins. He either won, came second or fell off in pretty equal numbers. It was enough to gain his second world championship and he blitzed the opposition at the Dutch TT.

Freddie Spencer had won the 1985 title. The V-twin was V-4 YZR 500 sliced vertically down the middle, thus using two separate geared crankshafts. The angle between the cylinders was 60 degrees, with the front cylinder pointing forward and downwards. Cylinder porting layout was as on the reed valve TZs, main exhaust controlled by guillotine powervalve with two booster exhausts, four main transfers and two vestigial ports on the rear wall of the cylinder. The powervalve was not used by Wimmer and Lavado at all races throughout the season, only at the slower, more twisty, circuits. A joined pair of 35mm flat-slide Mikuni carburettors were squeezed between the cylinders, feeding the six-petal crankcase reeds.

The crankcase itself was of small volume, following the trend of reducing crankcase size that had begun with the introduction of reed valves. The use of magnesium crankcases brought the bike under the 199lb (90kg) FIM minimum weight limit and there were also problems with big-end bearings turning in their crankcase housing. Aluminium crankcases were used to meet the weight limit, although they in turn suffered from cracking caused by the vibration developed by the engine. This also affected the standard aluminium TZ Deltabox frames used by the YZRs. Later V-twins were to be fitted with a balance shaft behind the engine to reduce the vibration. The dynamo of the Hitachi CDI ignition was

The Tenacious Twins

mounted on the left-hand side of the lower crankshaft, with the rotor on the upper crankshaft. The ignition fired both cylinders simultaneously. Side-loading gearbox facilitated the removal of the complete gearbox to make any changes to the gearing, although the characteristics of the engine made it less of a necessity than it had been in the past. With the exception of the cracked aluminium components the bike was very reliable and fast. All pole positions during 1986 were set by these machines and Lavado took his second world title despite crashing at Yugoslavia and Misano.

There was however one big problem with the V-twins, which affected Wimmer and Taira especially badly. They were very difficult to get started. Both riders got away at the back of the pack at most GPs throughout 1986, then used the power and handling of the bike to carve their way through the field to join the leading group halfway through the race. It made for spectacular racing, but it cost both riders a lot of GP points. The YZR500 riders had experienced the same problems a couple of years before, caused by the softer reeds that were being employed to enable the exhaust suction wave to draw fresh charge directly from the carburettor. The biggest challenge for those working on reed valve engines was to get the carburation right throughout the range of engine speed, without making it impossible to start. Also the simultaneous ignition of the cylinders required a really good 'bump' to start it, something Wimmer had lost a feeling for with the parallel twins that fired alternately.

During 1987, Kel Carruthers was also looking after the 250 machines and he had some ideas about improving the carburation:

> I bought some Motocross Keihin carburettors from the USA, made some manifolds and put 'em on. It was a different motorbike. It was typical Yamaha, 'Kel-San you can't do that.' First time we really run 'em was at Yugoslavia, when Martin (Wimmer) was riding with a broken foot. He was riding around in twelfth, but in the last couple

Wimmer's 1987 YZR250 was down on speed in its early season form shown here. Keihin flat-side carburettors fitted later in the year helped somewhat but it was a tough year for Yamaha.

of laps he was passing everyone and he finished fourth. In those days the Keihin were superior to the Mikuni.

TZ250T

Recognizing the inadequacies of the round-slide Mikunis still being fitted to the TZ250, the 1987 TZ250T was provided with a pair of state-of-the-art flat-slide 38mm TM Mikunis. These fed the ever-widening soft six-petal reed valves now measuring 72mm in width. A new set of exhausts and the TZ250T was ready for the punters. Clearly the minimum of development work had been done on the TZ, but unfortunately this was also true of the YZR 250. So overwhelming had been their domination during 1986, Yamaha might have thought it unnecessary to do extensive development work on the bike. The only change was a re-work of the ignition to bring rotor and dynamo together on the left-hand end of the lower crankshaft. The YZR was still fast and reliable, it had even been cajoled into starting well, but the NSR Hondas were better. Toni Mang in particular, working with his old friend Sepp Schlogel, got the Honda dialled in to not only produce excellent power, but to handle well. Forced to ride over the limit to compensate for the lack of power, Yamaha riders were frequent fallers in the GPs, both Lavado and Wimmer missing several races and being very de-tuned at others. The YZRs were passed out to other favoured sons at Sonauto and Ducados, but Frenchman Patrick Igoa and Spaniard Juan Garriga were no more successful than the veterans of the Yamaha team. One other new face appeared in the 250 class riding Yamahas and did well to secure three second places during the year. Luca Cadallora had taken the 125 Garelli to a world championship win in 1986 and stepped up a class. Many paddock watchers predicted a bright future for him.

The only glimmer of hope in the darkness of failure came from the USA where a legendary partnership had come into being. Bud Aksland, one of the foremost US tuners of Yamaha motocross machines showed that he knew a thing or two about twins as well and he found the perfect match for his tuning skills in one John Kocinski. Under the auspices and guidance from the king himself, Kenny Roberts, they took the 250 title in the US despite strong challenges from ex-GP stars Kork Ballington and Alan Carter, something they went on to repeat in 1988 and 1989, before taking on the rest of the world on the GP trail in 1990. Kocinski was nineteen at the time.

TZ250U

Back in Japan, there was considerable disagreement within the development department about the road that should be taken in the development of the TZ250. Despite a poor 1987 season, the V-twin had shown its mettle by taking the world title in 1986. There was a camp that thought that the higher costs of production that a V-twin would incur were the investment that was needed to get the TZ back to International race winning level. The other camp disagreed. They felt that there was still life in the old dog yet. Pointing to Kocinski's success in the USA, they argued that with some changes to the engine design they would be able once again to bring the TZ up to scratch, without the expense of tooling up for an assembly line of V-twins. Not only did they talk convincingly of the TZ winning again at national and international level, but they were prepared to back it for a glorious return to the GP trail. Their arguments were persuasive and the 1988 TZ250U was again a parallel twin.

This TZ twin was radically different from its predecessors. The most obvious change to

The Tenacious Twins

meet the eye was the fact that the cylinders had been turned, with the flat slide carburettors facing forward and the exhaust pipes stretching straight as a die to the rear of the bike. As well as the advantage of the straight exhausts, the inlet side benefited from the cooler air being used by the carburettors, mounted low and well away from the radiator. The cylinders were tilted very far forward, at about 60 degrees to the vertical to facilitate inlet and exhaust designs as well as keeping the mass of the engine low and forward in the frame. The crankcase was no longer split horizontally, but was a single casting with a sideloader gearbox as used by the YZR 250. Removal of the clutch and half-a-dozen retaining screws would enable the complete gearbox including shift drum and selector forks to be removed. Although the cylinders had been turned, the engine still ran backwards and the power was tapped from the centre of the crankshaft via the countershaft to the clutch.

More was changed than immediately met the eye. The direction of flow through the cooling system was reversed, with the crankcase getting the coolest fluid, with the flow through the cylinders and heads back to the enormous radiator. The previous design had over-cooled the heads, robbing the flame train of energy that would otherwise have been passed to the falling piston.

Compared to the engine, the chassis changed very little, with three notable exceptions. The triangulation tubing that had previously been above the rear swing arm, was now below, helping to get the weight low. The aluminium box between the twin frame spars behind the engine was enlarged considerably and now had the top of the rear suspension damper bolted to it. Also the TZ250U was the first TZ to have double iron discs mounted on the front wheel. Although heavier, the TZ needed the extra braking power provided by the pair of four-piston calipers.

True to their word, Martin Wimmer, Manfred Herweh and Jean-Phillipe Ruggia were provided with TZ250U machines for the GPs and offered a number of factory special tidbits, such as an electronically controlled Powerjet in the inlet of the carburettors. The control unit of the Powerjet system could be adjusted to change the engine speed at which it kicked in. The semi-works nature of these teams meant that they were free to go

In 1988 Yamaha supported both V-twin and parallel twin works efforts to decide the direction of the TZ series. Even with Helmut Fath looking after his machine, Martin Wimmer couldn't get his 250U to within spitting distance of a rostrum postion.

wherever they wanted for parts and Wimmer and Ruggia ended up using Hummel cylinders for most of the season, sometimes with and sometimes without powervalve. In addition Wimmer had managed to shave 22lb (10kg) off the 234lb (106kg) of the standard TZ, through the extensive use of titanium and carbon fibre for the seat. Of the three, Ruggia had the best season, managing a wonderful third place at the Spanish GP, but it was clear that the YZR riders had the better machine. Juan Garriga, Luca Cadalora and Carlos Lavado ran the V-twins and Garriga pushed title winner Sito Pons to the last GP, but had to settle for runner-up in the title race. In total the YZR won five of the fifteen GPs in 1988, the machine largely unchanged from the previous season. The 60 degree 'V' between the cylinders was opened up slightly to 70 degrees to improve the shape of the inlet tract. Flat-top pistons appeared as the factory experimented with flow angles of the transfer ports in the search for the ultimate scavenging configuration. It was only in the subsequent year that the 500s started using these pistons. A mixture of chassis were used, some riders sticking to the 1987 version, others using the 250U design.

One interesting development during the year was Kocinski's European debut race at Imola, riding a TZ250U with a carbon composite frame with the same geometry as the standard Deltabox. Built by one of the many British companies serving Formula 1 car racing teams, the frame would have brought the weight down to a value close to the FIM minimum of 199lb (90kg). Unfortunately Kocinski was taken out at the first corner by a late breaking Alberto Puig, and the subsequent GP at a wet Nürburgring also resulted in a crash. John returned to the USA to continue his total domination of the 250 class there on the TZ with standard Deltabox frame.

The first of many. Cadalora is ecstatic at winning his first 250 GP at the Nürburging in 1988. A good result, but the Marlboro Yamahas were overshadowed by the Ducados team with Juan Garriga.

TZ250W

The 1989, TZ250W was a fine example of minor evolution of the previous year's design. Externally, except for the bright red alloy wheels, the engine and chassis appeared almost identical. Sharp eyes would have picked out the new Nissin brake calipers front and rear, and might have noticed the rather fuller exhaust pipes a full 60mm shorter than the previous year. The Powerjet on the front of the TM Mikuni carburettors was now wired up to an

The Tenacious Twins

Despite getting points in all fifteen GPs in 1988, Juan Garriga couldn't prevent Sito Pons taking the title by ten points. Here at Donnington he won the battle for third place behind the backs of Cadalora and Dominique Sarron.

During the 1988 season Kenny Roberts was again pushing at the frontiers of 250 frame technology with this carbon fibre bike they built for a youngster from Arkansas who showed some promise. Jon Kocinski got to ride it twice before returning to the States.

electrically driven solenoid that kept the jet dormant until the engine reached 9,500rpm. No alternative chips were available to owners of the 250W wishing to adjust the range over which the Powerjet was activated.

Less obvious changes within the engine included a slight strengthening of the crank assembly and the usual game of 'hunt the horses' reflected in the changes made to port sizes, transfer duct geometry and piston skirt length. The front forks now featured the capability of damping adjustment. Fine tuning of the rear suspension completed the package for 1989. Carbon fibre dampers on the exhaust pipes and the use of Nissin racing calipers finished off the package.

The resulting bike was good, good enough to eclipse the Honda RS250 at national and international level. The Rotax teams had gradually disappeared from the international race meetings, with the noticeable exception of Aprilia, who dominated the European championships on essentially factory machines. Neither the TZs nor the Aprilias were competitive at GP level, where the battle was fought between YZRs and NSRs, the latter cleaning up with the first four places in the final championship positions. Sito Pons was at the top of his form, finishing no lower than fourth in any of the fourteen GPs. John Kocinski won both of the two GPs he entered and Luca Cadalora managed two more, but these were the only bright spots in an otherwise poor season. Jean-Phillipe Ruggia rode quite consistently until he dislocated his hip in a crash at Donnington resulting in a premature end to his 1989 season. In the USA, Kocinski easily took his third US 250 title despite breaking his wrist and missing a race as a consequence. His US hat-trick, the two GP wins, the inconsistency of other YZR riders, Kocinski's own desire to move on and the backing of Kenny Roberts all led to the simple conclusion that Kocinski would be campaigning the GPs for 1990.

In what was almost becoming a tradition, the Yamahas were significantly slower than the Hondas. Attempts to find some of the missing top end resulted in a terrible mid-season patch of unreliability for almost all the YZRs, with persistent detonation and crankshaft failures. Until 1989, separate crankshafts had been used, but now a single crankshaft was used as on the Hondas. Both 250 and 500 factory teams had problems with the crankshafts during the year. The problem was traced to the assembly that was now being incorrectly performed by a new department in the factory.

RACING IN THE 1990s

YZR250

There were three factory YZR250 bikes raced in 1990, Luca Cadalora and Alex Criville joining Kocinski. The bike had undergone a quite major re-design for the new season, most obviously identified by the change to a vertical left-hand cylinder and horizontal right-hand cylinder. In hindsight we can see that they were already preparing the bike for 'mass production' the following year, but it still differed significantly from the TZ unit it inspired. Both carburettors faced forward shrouded by a sheet of pvc intended to function as a still-air box for the carburettors to be fed cool air, free of turbulence. The enormous radiator just above the carburettors was curved to maximize surface area within the restricted width of the small engine. Upside-down Öhlins front forks were fitted to Kocinski's machine, but Cadalora used conventional Kyaba forks all season. Another thing Kocinski had and Cadalora didn't have was Bud Aksland looking after the engine. At the season kick-off GP in Japan, it became apparent that the Hondas of Spaniard Cardus and Dutchman

The Tenacious Twins

1989 was not a good year for Yamaha. Cadalora was the best of the YZR riders, with wins in Spain and Brazil, but plenty of crashes elsewhere.

Zeelenberg had a slight top-end advantage over the YZR. A weeks work on Aksland's dyno in California prior to the US GP, found enough top-end to remove most of the NSR's advantage. Kocinski didn't seem to mind one way or the other; he rose to the challenge of steering rather than powering the YZR to his first world title. It wasn't all plain sailing and Cardus in particular pulled a midseason points lead over Kocinski after crashes in France and the UK. Kocinski, going into the last race with a five-point deficit, concentrated on the win, while Cardus totally blew his cool when the gearchange lever snapped towards the end of the race. Cadalora managed a good third place in the final standings with wins in Japan, Austria and the UK. Honda's three-year hegemony had at last been broken and the YZR had proven reliable and balanced enough to take the title. It was time to make it available to the masses.

The Tenacious Twins

Suberb as Jon Kocinski was during his first GP season in 1990, he messed up a few times. Here at the German Nürburgring he ran wide on the last corner and let Dutchman Wilco Zeelenberg through to his first and only GP victory.

The Last of the Parallel Twins

Considering it was to be the last of the parallel twins, the TZ250A available for the 1990 season was significantly different from the W. This suggests that the decision to move to the V-twins from 1991 had not yet been taken. The most significant changes were the smaller crankcases, some 3cm narrower than before, a tighter water jacket round the cylinders, with a more powerful water pump, and a return to base-mounted cylinders.

The best single result of 1990 was probably the win in the 250 Junior race at the notorious Isle of Man. Privateer Ian Lougher beat Steve Hislop on a kitted RS250 to win the 150 mile (242km) race at an average speed of 115mph (185km/h). In the European Championships Dutchman Patrick van den Goorbergh, finished third with the same points as Aprilia riders van der Heyden and Pennese, but with no outright wins. The parallel twin was still a competitive machine, still had the potential to mix it with the Hondas and Aprilias for a few years, but was becoming too expensive to produce. The street-based engines that had been so close in design to the TZs in the 1970s and 1980s, were gone, the late 1980s TZR250 being the last of the breed. There was still a healthy market for race bike replicas, especially in the Japanese domestic market, and after

The Tenacious Twins

Although well out of it on the GP trail, the TZs were still holding their own at National level. Patrick van den Goorbergh managed third place in the 1990 European Championships on his TZ250A.

Kocinski's 1990 world title, a replica was guaranteed to sell well. By sharing some of the tooling costs, it would be possible to sell the production TZs for a price that would keep racing accessible to the common man. The 1991 TZ250B could be offered for sale at a price only slightly higher than the TZ250L seven years before. But what a difference in sophistication between the two.

The V-twins

As had been debated more than five years before when the first works YZR V-twins appeared, this configuration offered two distinct advantages over a parallel twin. Firstly the engine could be made considerably narrower without compromising water jacket thickness or impinging on the dimensions of the transfer ports. This had not been a design requirement of the old engine, with the drive to the jackshaft behind the engine being taken off the centre of the crankshaft. Although using a single crankshaft in contrast to the early works YZRs, the drive had moved from the centre to the right-hand end of the crank.

The second potential advantage lay in the inherently smoother delivery of power. The two in-board flywheels were very narrow, bringing the plane of the two cylinders close, but not entirely in-line. If they had been exactly in line, the 90 degree 'V' of the cylinders would have offered an almost perfectly smooth power delivery. In order to cancel the slight rocking motion that would otherwise be present, a second gear on the right-hand end of the crank drove a balance shaft behind the engine. Now with a very smooth

The Tenacious Twins

power delivery, it was possible to bolt the engine directly into the Deltabox frame, without the danger of cracks appearing in the frame as had happened not infrequently on the old parallel twins.

Some of the old features of the TZ were still around, including guillotine exhaust valves and Mikuni 38mm TM flat slide carburettors. A lot was new. The cooling system had been re-designed, to flow water faster and prevent eddying caused by sharp changes of direction of the water flow. The port timing was significantly different. The exhaust port height was lowered 1.5mm and was 1.0mm narrower. The transfer port nearest the exhaust moved towards the gaping port and its roof was radiused down in a smooth curve to the floor of the port. This wall of the port also had a lip just at the exit into the cylinder, which turned the flow of the charge 40–45 degrees towards the rear of the cylinder, considerably sharper than had been used on previous TZs. The secondary transfers grew in width by 4mm but were of conventional shape. The total effect of the changes in port timing were to improve the mid-range of the engine at the expense of top-end speed. The power band came in at 9,500rpm and signed off at 12,000rpm.

The engine was now turning forward again after running backwards for ten years and driving a jackshaft to correct the direction. It was no longer necessary to run the engine backwards as the large inlet port that had destroyed the pistons of the TZ250G was gone. Crankcase reed valves had caused the inlet side to become populated with relatively small transfer ports that supported

The V-twin heart of the TZ250B. The YZs on which it was based had the upper cylinder turned 180 degrees for better exhaust pipe design. Note the powervalve bulge over the exhaust port.

Initial YZR250s had used separate crankshafts, truly reflecting their 'half-a-500' heritage. By the time the TZ250B went into production a single crankshaft was used.

143

The Tenacious Twins

the piston skirt quite adequately. The change, could and perhaps should, have been made much earlier. Gearboxes had received almost no attention in the last decade once a cassette system had been introduced. Now a refinement arrived in the form of 'stacking' of the two gearshafts. The mainshaft driven by the clutch was located forward and above the driveshaft. In the drive to produce as compact an engine as possible this saved another couple of centimetres.

The compacter engine of the TZ250B was just another step in the move to distribute the weight of the bike to load the front tyre more heavily. The more weight on the front wheel the more stable the bike becomes, permitting more radical rake angles to help the bike turn quickly in the corners. Improvements in tyre technology had made this possible without the danger of unloading the rear wheel so much that it would spin rather than drive the bike forward. The 250B chassis was of the now commonplace Deltabox design, although Yamaha did not consider it important enough to give it a sticker proclaiming itself as such. There were no longer frame tubes running under the engine, clamped as it was between two flanges extending down from the twin spars. It was now a fully fledged stressed member of the chassis. Looking at the left-hand side of the bike, knowledgeable TZ observers would have done a double-take when they saw the new swing-arm. The design seemed almost identical to that last seen on the 250T four years before. The triangulation had now swapped back to being placed above the swing-arm, with massive box gusseting between the upper and lower halves. On stepping round to the other side of the chassis, it would become clear that the swing-arm was totally different with a 'gull-arm' design being used to clear the exhaust pipe sweeping back from the horizontal cylinder. Detailed changes to the rear suspension were in contrast to the totally new upside-down fork used on the front. It would have been a shame to compromise all the newly found rigidity of the chassis by fitting inherently sloppier conventional front forks.

This technological masterpiece retailed for $15,000 in the USA and $20,000 in Europe, rivalling but not surpassing that all time bargain the TZ250C. Unfortunately it wasn't perfect and a bulletin was quickly issued warning of over-tightened engine bolts and suggesting the removal of 0.7mm from the height of the exhaust port and the widening of both main and auxiliary exhaust ports. For the GPs, Spaniard Alberto Puig and Italian Paolo Casoli received special kits, which included six-port cylinders and cylindrical exhaust valves as used on the YZR500. This was the most the factory would do in the 250 class. They were too busy with the YZR500 and the plans to build a batch of machines for 'over-the-counter' sale. It didn't help much and the GPs were dominated by Aprilia and NSR Hondas. Aprilias ran away with the European titles, but in British and USA 250 racing, with a few modifications, it was a very successful machine.

TZ improvements

Despite all the work involved in producing their first batch of production 500cc racing engines for ten years, Yamaha found time to give the TZ the once over for 1992. The six-port cylinders found their way onto the TZ250D from the 1991 factory kit, but the cylindrical powervalve did not. It was quite a tight squeeze to accommodate all six transfer ports plus the single finger port opposite the exhaust port. The transfer port closest to the exhaust pushed up even closer under the curved walls of the trapezoidal shaped port. It also lost some of its backflow angle so that the charge was not directed so far back in the combustion chamber. Raising and widening

the UK, USA and other countries. Colin Edwards II was the man to beat in the USA and few managed it. Steve Sawford took the British Championship after two other 250D-mounted challengers for the title crashed out of the last race of the year. The 1992 Isle of Man TT was another Yamaha benefit with only one Honda in the first fourteen places. Brian Reid won the race bringing Yamaha their hundredth Isle of Man TT win.

Yamaha felt that the 250D had been strong enough not to warrant significant expenditure of time and resources in major development. The 1993 250E appeared with slight changes to the transfer and auxiliary exhaust ports as well as revised mapping sequences of the digital ignition. A 0.25in wider front wheel rim and fairing and set giving better aerodynamic flow completed the update accorded to the TZ. At $16,500 in the USA and $20,000 in Europe, they were definitely good deals for 250 national riders. Around the world it did well for itself, taking national titles in the USA, Holland, France and Belgium, narrowly losing out to Honda in the UK and seriously swamped by a big Honda effort in Germany.

On the GP trail, the TZ-M was generally voted the machine least likely to win the championship in 1993. The Aprilia versus Honda battles of 1992 were expected to continue with Aprilia having the upper hand over Honda. Suzuki had stimulated Jon Kocinski their new rider, into performing

In 1993, Schmidt's TZ was claimed to be the same as that run by title winner Harrada, but there was a world of difference in the results.

some startlingly fast test sessions during the winter, leading to many expecting a three-way battle between these manufacturers. Nothing was heard of Yamaha's plans throughout the winter. Jochen Schmidt was to continue to get support and Pier-Francesco Chilli had stuck with his South African sponsors Telkor, who would also be getting factory equipment. Yamaha's agreement with Telkor was to supply machines for a two-man team, consisting of Chilli and Tetsuya Harada. Telkor accepted Harada as a means to getting factory machines for Chilli. They couldn't have realized that Yamaha accepted Chilli as a means of getting sponsorship money for Harada.

Few people had heard of Harada as the season started at the kick-off GP in Australia. He'd come second in the 1992 Japanese GP and sixth the year before that, but few Western journalists placed much store on the results of Japanese riders at their home GP. In Australia he beat Jon Kocinski by 0.003 seconds, causing a few heads to turn in the process. In Malaysia he was second to Japanese rival Aoki on an NSR Honda, but with the subsequent win in Japan, it was clear that Harada could mix it with the best. Still some put his performance down to the 'home game' effect of racing in the Far East, until he took the first European GP at Jerez in Spain. It was clear that his astounding performance was more due to himself than the machine, as none of the other Yamaha riders came even close to matching his performance. The Yamaha was not fast, down by at least 6mph (10km/h) on the Hondas, but, as tradition would have it, had excellent steering characteristics. Harada had been riding the TZ-M for two previous years in the Japanese national championships and was totally familiar with the machine and at the top of his form.

The GP circus moved on to the fast Germanic tracks at Salzburg and Hockenheim and mid-field places were the result. At the technically demanding circuits of Assen and San Marino, he was back on the rostrum, looking like the runaway title winner. Then it all seemed to turn sour, when a high-side at the first corner of the British GP left him with a fractured shoulder and a forty-five point lead that could never hold off an on-form Loris Capirossi. By the start of the last GP of the year it looked like the Italian had done it with a ten-point lead in the series. In the event, Harada achieved the impossible, with a superbly calculated race to the front of the field with just two laps remaining and Capirossi's tyres causing lurid slides and allowing him no better than fifth place at the chequered flag. Harada was only Japan's second world champion, 16 years after Takazumi Katayama took the 350 title ... on a TZ Yamaha.

9 The Doldrums
Factory 500s 1981–3

Grand Prix racing during the 1970s had seen all classes come together in terms of the relative numbers of factory and private machines being entered. There was always a handful of factory specials being used by the chosen ones, but at least 90 per cent of the field was made up of totally dedicated racing individuals, some with better sponsorship deals than others, running machines available 'over-the-counter' for anyone with the cash. The factory riders usually won, not only as a result of better machinery, but also their own skill reflected by the fact that the manufacturers had entrusted them with the technological masterpieces for the GPs.

Factory riders, though, did not have a monopoly on talent, and part of the magic of the sport came from the 'what if' scenarios, bringing together privateers and works specials and endlessly discussed by enthusiasts throughout the world.

The 500cc class had become a fascinating class to watch from the middle 1970s, the decline of MV being matched by the rise of both Yamaha and Suzuki. But it wasn't so much the factory 500s from the Japanese companies that had rejuvenated the class, but the arrival of competitive full capacity production race machinery. In particular the Suzuki RG500 narrowed the gap between

Jack Middelburg starts the run into Woodcote corner at Silverstone at the 1981 British GP. Fifty metres later he took the flag to become the last privateer to win a 500cc GP.

factory and private rider to a level that made it almost conceivable that the underdog might win. The few times that the impossible became reality, and the factory teams were beaten, there were mitigating circumstances, usually weather-related. The one exception was the memorable British GP of 1981, the high point of Dutchman Jack Middelburg's career and possibly the high point of the 'production racer' *per se*. Middelburg riding a Mk VI RG Suzuki beat Kenny Roberts at Silverstone in a straight race to the line. Twelve years and more than 100 races later, his place in history as the last privateer winner of a 500cc GP race is unchanged. Throughout the 1980s, the gap was to widen again to unbridgeable proportions. The 500 class was to become the playground of the factory teams in their determination to achieve victory at any cost.

During the second half of the 1980 season, it became clear that the works Suzukis were faster than the OW48 in-line piston-ported four-cylinder. Roberts was pushing the factory to come up with a quantum jump in performance for the 1981 season, to bring him back into contention for what would be his fourth world championship title. Yamaha saw no way to achieve this using the same parallel four-cylinder layout that they had been using since 1973. Not only was the engine too wide in this form, but it prevented the use of the disc valve induction that was so fashionable at the time. By the end of the 1970s, all GP machines in all classes were fitted with disc valve induction with the exception of the Yamahas.

OW54

Disc valves had been attractive to two-stroke engine designers from the early MZ days of the 1950s as they provided a simple mechanical way of providing asymmetrical inlet timing. The inlet tract could open early and close as soon as necessary to prevent blowback into the carburettor. The classic piston port engine required complex analysis and tuning of the inlet tract to achieve the same effect via the standing wave between the inlet port and the mouth of the carburettor. Reed valves fitted to piston port engines, could prevent blowback at the expense of reduced maximum airflow, i.e. lower maximum power. Disc valves seemed to be the way to go for the factory in their no-holds barred fight to retain the world championship title. Suzuki had developed what seemed to be the best 500cc engine of the era; Yamaha would take the design and beat Suzuki at their own game. Yamaha built Roberts a 500cc square-four for the 1981 season and called it the OW54.

Engine layout was identical to the Suzuki, not being a true square four like the one Suzuki had campaigned with miserable results back in the 1960s. To avoid too long a wheelbase, the rear cylinder pair was mounted above and behind the front pair, and tipped 30 degrees forward. Convinced of its effectiveness, Yamaha retained the powervalve system they had been using for the previous three years. The cylindrical version had now been replaced by a guillotine plate that was lowered and raised by an electric motor. The Suzukis had no variable height exhaust port system and also differed by having a total of seven transfer ports in comparison to the five found on the OW54. The engine was squeezed into an aluminium frame of the same dimensions as the one used in 1980, although it was no longer camouflaged by black paint. The chassis design was essentially unchanged, with the traditional monoshock rear suspension passing over the top of the engine.

Although the racing world was expecting a new OW from Yamaha, few details had escaped from the research and development department during the winter of 1980/81. The whole paddock turned out to see how it

The Doldrums

The OW54 Yamaha RG500 replica that marked the start of the three year period that saw Yamaha YZR500s in the doldrums.

would do when it was first wheeled out for practice at the Austrian GP in April. It was not an auspicious debut as Kenny described in his book:

> The first time I rode it was at the Austrian GP at the Salzbürgring in '81, and I did about sixteen laps to break it in. All they had was one complete engine and a spare set of crankcases. After it was run in, they said I could go for it, but there was no way it would run. I came back in and said there was something wrong, either it was running on three cylinders, or if it was running on four they had better put it in the crate and send it back to Japan.
>
> They changed the carburation but it still would not run. They had this bike that was supposed to turn out about 150bhp, and it was a pretty big engine because they had no idea what that much power would do to the transmission. We raced the square four and it never got any better. It just didn't get any better. It had a funny sort of power delivery. It ran as though it was never going to take off and reminded me of an engine which had the transfer ports too high. Anyway the shock spring broke in the race so that saved us the trouble of having to finish it.
>
> We went to Hockenheim for the next weekend and I had a meeting with the Yamaha engineer, Doi, and Kel and Mike (Maekawa). I told them that I thought the transfer ports were too high. Doi said that was impossible. He said that the factory would check it out and I replied, 'Hell we have to race at the weekend'. Never mind waiting for the factory to check it out. I wanted Kel to machine the base of the cylinders. They threw a fit. We only had eight cylinders and two of them had seized in Austria, because they were trying to lean it out so much to make the engine run – not to mention that it had thrown me off.
>
> There were some real intense meetings. As often as I would tell Kel to do it, Doi would tell him not to, until I finally said, 'Kel, go and machine the cylinders'. Doi was still screaming 'No, we must test it in Japan', but it would have taken three days and been too late. When he walked into the truck, Kel was working at it, the lathe was turning and bits of swarf and cylinder were flying everywhere. He just went red and walked out.
>
> We won that race, we just smoked 'em. We discovered that the castings were way out, half a millimetre too high on the

The Doldrums

Kenny Roberts took some time to adapt to the change from the pushing power of the piston-port engines to the light switch delivery of the disc valve OW54.

transfer ports, and we won two races in a row.

Not only were there quality control problems with the OW54, but Roberts had difficulty adapting to the power delivery of a disc valve engine. Although having a wider power band than the OW48, it was not a smooth delivery; either nothing or tyre spinning power as the power band was reached. This deficiency of disc valve engines is caused by the weaker carburation metering signal as the disc valve opens at low engine speeds. The faster the inlet tract opens, the better the resulting atomization of the fuel. In the power band, not only is the piston travelling faster, but the exhaust pipe suction makes a significant contribution to the metering signal.

Roberts won in Germany and again in Italy, but was far from happy with the handling of the machine. The fast circuit of Hockenheim and the rain in Italy enabled Roberts to ride competitively, but the French GP was a disaster despite a new frame with the engine lower and further forward. The 6in Goodyear rear tyre overheated and Roberts dropped back to fifth place after leading for the first three laps. In Yugoslavia, the hardest compound tyres were shot by half distance and Roberts settled for third place. The world title starting

The win at the West German GP at Hockenheim made it look as if Roberts could retain his three year old 500cc crown. The rest of the season was a huge disappointment.

to slip away by Yugoslavia, disappeared out of sight after the debacle at the Dutch GP when a brake pad was inserted back to front and fused itself to a front disc rotor during the warm-up lap. The new frame, carbon fibre forks and special triple clamps Roberts had received at Assen were put to better use at the Belgian GP, where he was narrowly beaten by Marco Lucchinelli on a Suzuki Gamma. The carbon fibre sliders in the forks suffered initially from too much stiction; when they re-appeared later in the 1980s, they were coated with a thin film of chrome and that eased the problem.

Roberts spent the San Remo GP at Imola flat on his back recovering from food poisoning and at Silverstone he lost the duel with Middelburg. The last two races of the year were the Scandinavians, headed up by Imatra, the infamous Finnish street circuit, where a powervalve failure meant that Roberts could do no better than seventh place. A titanium screw on one of the linkages for the powervalves broke and the valve on that cylinder stuck in a closed position, giving the impression of a partial seizure. Roberts was philosophical after the race, saying 'You can't really blame Yamaha for the powervalve breaking, they don't have a railway crossing in their test track'. The final GP of the season was a good result for Yamaha, with Barry Sheene, also riding the OW54 since the French GP, taking the win and Boet van Dulmen bringing home his TZ500J in second place. Roberts packed it in after an incorrect tyre choice. Plagued with tyre problems throughout the season, Goodyear withdrew from the sport at the end of 1981, never to be seen again.

There is a theory that Yamaha never really intended to produce the OW54 as a square-four but, under pressure from Roberts, produced it as a stop-gap while they prepared their preferred machine, the V-4 OW61 of 1982. It was comparatively simple to produce, since Suzuki RG 500s were easily obtainable and had appeared reasonably simple to 'improve' during dyno testing. The problem had been the incompatibilty of the resulting power unit with Yamaha's monoshock chassis, Goodyear tyres and Roberts riding style. The Suzukis had a quite smooth flow into the power band, whereas the Yamaha had a step-like transition. During the winter of 1981/82, some more development work was done on the square four smoothing out the power delivery, building a more compact engine block and perhaps most important of all, building a new chassis for Graeme Crosby

The Doldrums

and Barry Sheene who were to ride the OW60 square four for the complete 1982 season, guarding Roberts' back.

Crosby's transfer from Suzuki to the Yamaha team run by Giacomo Agostini with sponsorship from Marlboro, was of tremendous assistance to a factory trying to build a clone of their competitor's bike. Crosby tested Roberts' OW54 at the Yamaha test track at Iwata and declared it impossible to ride, his times being six seconds under the lap record. Roberts had tried to compensate the poor turning ability of the OW54 by tightening the steering head angle to a radical 22 degrees, then used triple clamps machined at an angle to help reduce the drop in trail that occurs when riding over bumps. Roberts described this as getting the wheel to 'roll over bumps'. Crosby (and Sheene) were front wheel steerers and needed more conventional rake angles to turn the bike comfortably. Crozo had a simple solution to the problem in the form of the dimensions of the Suzuki Gamma frame he had been riding the year before. With the new frame Crosby was on the lap record and Sheene knew enough about Crosby's riding style to realize that what was good for Crosby would be good for him.

OW60

Crosby, Sheene and Marc Fontan all received OW60s for the complete 1982 season, and Kenny Roberts used it for the early Argentinian GP and Daytona. He may later have wished he stuck with it all year. The

Kenny's ride on the OW60 at Daytona should have convinced him to use it for the rest of the season. He might have been world champion if he had.

first race of the year was Daytona, with Roberts taking pole position on the OW60, while Crosby made do with one of Roberts' old TZ750s. The OW60 ground to a halt after a dozen laps with a broken crankshaft bearing, and Spencer and Mike Baldwin, riding big capacity Honda four-strokes, kept on disappearing into the pits to check out their overheating Michelin tyres. Eddie Lawson, riding superbikes for Kawasaki in the US at the time, was entrusted with their disappointingly unsuccessful GP square four for the race, but retired with a broken gearbox. As riders and machines in increasing numbers collected in the pits, Crozo and the 'obsolete' TZ750 droned on and on, until the flag went out and Crosby had his first 200 miler win and the TZ750 its ninth consecutive win.

Down to Argentina and a win for Roberts on the OW60 which he claimed was significantly better than the OW54, the chassis and suspension working well. Sheene was a close second with Freddie Spencer giving Honda's first two-stroke GP road racer, the three-cylinder NS500, a tremendous debut just behind. Crosby's OW54 had also broken a main bearing. After the race, there was a mad dash by the Italian riders and teams to make it back to Imola for the next weekend's 200 miler. Although losing some of its glamour, there were still three factory teams represented, Franco Uncini riding Suzuki Gamma, Marco Lucchinelli riding the Honda RS500 and Crosby riding the OW60. Trevor Tilbury was looking after Crosby's bikes at the time and he remembers the race well:

> We made a perfect pit-stop in the first leg of the race, when Uncini overshot his crew and Lucchinelli replaced the Honda battery. Crosby won the leg, but the bike had been detonating badly.
>
> There was only about 45 minutes between the heats, so I pulled the exhaust pipes off and you could see it was detonating around the exhaust ports. This was unheard off on the square fours. We filed the piston very carefully through the exhaust port. It was the two front cylinders that were doing it, so we got under the bike and filed it away. Cleaned it out with compressed air.
>
> Then Crosby went out and won the second race also. But it was tactics that won us the second race. We'd planned the gas stop for the first race after twenty-one, twenty-two laps. Crosby made a good start and Lucchinelli was coming. I went to Agostini and said 'Ago, we're going to change the plan, because they know when we're going to gas stop and they're going to go for the same time'. We pulled him in laps early. Lucchinelli was going to make his big move at the grandstand. He was going to nail Crosby at the chicane before the pits. He came up the inside of Crosby watching the brakes, just as Crosby put out his right foot, signalling his intention of pitting. Crosby never stopped, just shot into the pit lane. This totally messed up Lucchinelli. He should have won the race but he screwed his brakes up and went into the tyre wall. By the time he'd pulled himself out, we'd gassed up Crosby and sent him back out before Lucchinelli came by. Lucchinelli still had to stop for gas, so we won. They were mad as hell. Agostini was up in the clouds.

The carburation was finally sorted when Mikuni provided Agostini's team with a set of needles and jets and Agostini hired Mugello for three days of testing. Tilbury and 'Radar' McCullen, also working in the team, managed to keep the settings secret for a time but were eventually asked by Yamaha to pass them on to the other OW60 riders in mid-season.

In the hands of Sheene and Crosby the OW60 was a good machine, better than the OW61 that Kenny Roberts was wrestling around the circuits all year, but not quite as good as the Gamma in the hands of Franco Uncini. It would have been a second place for Sheene and a third place for Crosby if

The Doldrums

All the main contenders lining up for La Source at Spa in Belgium 1982. Middelburg (4), Sheen (7), Crosby (5), Roberts (3) and Spencer (40) use the traditional line. Lucchinelli (1) thinks he's got a new line worked out.

Sheene had not ploughed into a fallen machine during unofficial practice at Silverstone during August. Two badly broken legs and a fractured fore-arm ended his 1982 season and nearly his racing career, and Crosby took second place in the final results and Roberts and Sheene, the great rivals, fittingly ended up sharing fourth place behind Freddie Spencer.

OW61

For Roberts it had been his worst GP year ever, with tyre problems, chassis problems, engine problems and finally a big crash at Silverstone. The OW61 was not an easy machine to ride. Yamaha had adopted the 'V' rather than the square configuration to enable them to lower the centre of gravity of the bike, by placing the engine lower in the frame, without the risk of the carburettors grinding along the track. Just two disc valves, fed by two double-choke carburettors, were situated in the 'V' of the pairs of cylinders, requiring extra shafts to drive them at engine speed. It also made carburettor accessibility a problem and compromised the flow of clean cool air. The exhaust pipes on the forward cylinders also were problematic as, when correctly dimensioned for the engine, they dragged at the lean angles now being achieved on cornering. After much experimentation, Yamaha chose a conventional path for the pipes, but built a new frame which abandoned the cradle design of the past for one where the lower tubes passed along the side of the engine

The Doldrums

From bad to worse. The square four OW54 made way for the V-4 disc valve OW61. Carruthers calls it the worst Yamaha that Yamaha ever built.

rather than under it. Even this design of frame was not without its problems, as the tube on the right-hand side of the engine was forced to loop up and over the clutch housing to the main spar running from the top of the headstock to the back of the engine. At least the exhaust pipes now fitted snugly under the engine.

Another strange feature of the OW61 was the rear suspension design. After years of adhering to the monocross design, they went totally mad and introduced a horizontally mounted rear shock just behind the engine that was compressed at both ends via a bell crank that translated the vertical movement of the rear wheel into lateral pressure on the shock. Once again this was an attempt at lowering the overall centre of gravity of the machine, but continued to rely on variable thickness of the spring to provide rising rate of resistance to compression.

They had also implemented a weight saving exercise throughout the bike, including the engine, making it in Robert's words '... like a light switch, either on or off'. One of these weight reductions had involved the use of magnesium inserts in the crankshaft flywheels. Just as mind-bogglingly expensive was the use of titanium exhaust pipes on and off throughout the season. It turned out to be too light, spinning the enormous Dunlop rear tyre without hesitation, causing it to overheat and go off long before the race was over. Kel Carruthers believes it was a big mistake:

> The OW61 was the worst racing motorcycle Yamaha ever made. The OW60 was a good machine, much lighter than the over-engineered OW54, and with a decent chassis. There's no doubt Kenny would have been World Champion in 1982 if he had ridden that bike all year. But he got to hear of the V-4 and kicked up a stink about riding it, resulting in it arriving in Europe before it was ready. It had no bottom-end power.
>
> The suspension system was a disaster. It had so many different links in it, that as soon as there was a little bit of clearance in all the little joints, you could move the back wheel up and down without the suspension moving. It was maybe the worst Yamaha that Yamaha built.

157

The Doldrums

The chassis on the OW61 was a disaster, the horizontal rear shock absorber system had too many linkages and the 8in rear tyre totally dominated the motorcycle.

It was in these circumstances that Roberts talks of his victory at the Spanish GP on the tight twisty Jarama circuit as one of his greatest races. 'I had to ride it off the kerbs because the rear end would just light up and take off anyplace but where I wanted it to go'. His race season was only to last three more GPs after the Spanish, with disappointing fourth places in Italy and Belgium and a second place at a re-started Dutch TT. In Yugoslavia he retired with ignition failure. Despairing of getting the OW61 sorted, he tried the OW60 in practice at Silverstone, along with a V-4 engine without powervalves, but he was not satisfied with either machines. At the first corner of the race he crashed, breaking a finger, and retired for 1982 after so much disappointment. It was a year he would sooner forget.

There were not to be any Yamaha squarefours in GP racing in 1983, but they had not disappeared completely. Although Yamaha had won Daytona in 1982, it had been considered a source of some embarrassment that it had been the TZ750 of Graeme Crosby that had saved the factory's bacon. The OW60 of Kenny Roberts had failed, raising questions about the capabilities of the official factory machine. Yamaha were determined not to let that happen again and

158

built a Daytona special to ensure it. The OW69 was based on the OW60, but had special crankcases, cylinders, heads, cranks, pistons and exhausts developed to match the 64 × 54mm bore and stroke. An extra radiator was fitted slung down low in front of the forward cylinders. No significant changes were made to the chassis of the OW60. Displacing 695cc, it was thought at the time to develop around 150bhp, but just to be safe a fully-tuned version of the engine was also developed and brought along to the track with a reputed 175bhp available. It remained in the garage as Lawson and Roberts lapped Daytona spinning the rear tyre for much of the way. It all turned out to be enough as Roberts and Lawson took first and second places.

A year later the same bike was wheeled out again for the same riders, although some subtle changes had been made to make it slightly more ridable. The 1983 model had been powervalve-less, but now guillotine valves were fitted. Slightly rich jetting and retarding of the ignition were requested by Roberts after practice as he still considered it too powerful. Roberts again won the race, but was sharply critical of the mix of riders and machines that saw a 50mph (80km/h) speed difference between the fastest and slowest riders. Daytona and US road racing was going through an identity crisis that was to last for a few years to come.

OW70

For the GPs, it was the OW61 that underwent evolutionary change to become the OW70 and evolution rather than revolution was exactly what was needed. Some important changes were made to the chassis, with the large spar to the top of the headstock and the tube to the bottom becoming a single full length spar right back to the swingarm bearing behind the engine. The Deltabox frame was now only a minor update or two away. The tubes supporting the engine ran tighter round the engine than they had on the OW61, with the radiator no longer within the frame, but mounted on the outside of the frame tubes. Anticipating cooling problems, a small radiator was also mounted low down in front of the forward facing cylinders in the 'V' formed by the fairing belly pan. The frame also doubled as a tank for a couple of litres of fuel, something Kawasaki had done the previous year on their KR500. This was never actually run during GP races as the FIM had banned the practice during the winter of 1982.

There were not many changes to the engine, but enough to cause the Roberts team some real problems during practice for the first GP at Kyalami in South Africa. Kel Carruthers recalls the problems:

> The bikes wouldn't start. South Africa had high altitude and low pressure and we'd push it for 100 metres before we'd get the thing to start. Of all the mechanics and myself, I was the one that could start it the best and I ended up starting all the bikes all the time. Once started it had no mid-range power at all. Before the race I'd been down to a motorcycle shop in Jo'burg and re-machined the cylinder heads and raised the compression. One time I came back from down town, practice was ready to go and the two bikes were sitting there without the heads on. The mechanics threw the heads on and the riders went. It was a disaster really, but we got second in the race. I discovered eventually that the inlet timing was wrong. It was open too long and we could get no primary compression. For the rest of the year they sent me plain discs and I'd cut them myself. Once we'd sorted that out it ran good.

Two major changes took place with the suspension. Firstly the rear suspension system was re-designed and secondly Öhlins were

The Doldrums

Now we're getting somewhere. The tiny Öhlins sticker on the fairing of the OW70, represented a transformation of the machine. Suddenly it began to handle well.

brought in to take care of the rear damper. The horizontally mounted shock was gone, replaced by a design closely resembling the old monocross. The swing-arm itself was triangulated with a massive gusset between the upper and lower arms. The lower end of the damper was fixed to a lug in the middle of the gusset and the upper end to a cross brace just behind the steering head. The relationship with the Swedish suspension experts had been born from a need for expertise to cure problems the Yamaha motocross team had been experiencing. So successful were they, that they were asked to assist the road race team, and Roberts was ecstatic:

As soon as I started working with Öhlins they began talking low speed, medium and high on bump and the same on rebound. I am not a shock engineer but I could say, 'I think this one needs a little less rebound high on top because in fast corners it is just tying me down a bit' and it meant something to them. The first guy I worked with was Mike Mills and it was like heaven. For the first time I could dial the motor cycle in and it would stay there. It was great in '83.

Actually it was not until mid-season, when a new frame arrived with rear suspension with rising rate geometry, that Roberts really got the suspension sorted. The new design featured a near vertical shock passing through the massive gusset of the swing arm mounted at the top to a cross brace of the rear sub-frame and with bell crank linkages at the bottom to the bottom of the main frame spar. Available from the third round in Italy, it was debugged over the next four GPs. By Assen it was ready with a dramatic improvement in results. Eddie Lawson, now Roberts' number two rider, rode the monoshock design all year.

The front suspension had also undergone an update, the oil reservoir on the front of the lower front legs that had been run in 1981 and 1982 being replaced by a short finned unit. Just behind the fork leg was a mechanical anti-dive unit that changed the internal fork jetting as the brake caliper moved. Anti-dive systems of various designs had been tried by many teams during the early 1980s, each hoping to minimize the change in steering geometry that occurs under heavy braking and to prevent bottoming. None of them offered a perfect solution to the problem as they always resulted in blocked suspension, feeling to the rider like totally bottomed suspension. By the mid-1980s they had disappeared from GP bikes, and sophisticated spring windings were used to ease the problem.

Nineteen eighty-three was to be a classic season of GP racing in the 500 class, as Spencer and Roberts shared the GP wins between them. In the first half of the season, Spencer had the advantage, only for Roberts to gain an edge as the new rear suspension was perfected and Spencer started experiencing problems with his Michelin tyres. The most famous race of the year was the Swedish GP which featured a last lap 'do-or-die' overtaking manoeuvre by Spencer, which resulted in both riders taking to the dirt on the penultimate corner. Spencer, riding the lighter more manoeuvrable Honda triple recovered first to pass Roberts and take the win by 0.16 seconds, setting himself up to take the world title at the last race of the year at Imola, by two world championship points.

It was fitting that Imola should be the scene of the last GP race of 1983, as Kenny Roberts had declared his intention to retire as a GP rider at the end of the season. Imola had been the scene of his European debut in 1974 when he had been beaten by Agostini due to sloppy pit work and general inexperience. Although bitterly disappointed at missing the 1983 title by so close a margin, he was pleased to have won his last GP at this famous track, and was determined to put his enormous experience that he had developed in the intermediate ten years to good use within the sport. It was good to see such a talented rider quit (with the exception of a handful of appearances in 1984 and 1985) when he was still on top, instead of slipping slowly and painfully into obscurity. It was typical of the character of the man.

Roberts' disappearance from the GP grids coincided with developments within two-stroke engine technology that would result in the universal application of a common engine format between all the big three manufacturers competing in the 500 class. Honda's maverick reed valve triple, Suzuki's veteran disc valve square four and Yamaha's disc valve V-4 were to be swept aside, doomed to be consigned to the crusher. The reed valve V-4 had arrived in town and was here to stay.

The Doldrums

Roberts' last GP at Imola which he won but the championship went to second place Freddie Spencer.

Freddie Spencer has just become World Champion at the San Marino GP Imola, but you wouldn't know it. King Kenny savours his last GP win and surveys the crowd gathered at his feet. The end of a truly great career.

10 Biggest Bang per Buck
Factory Reed-Valve V-4s 1984–93

OW76 AND THE INTRODUCTION OF REED VALVE INDUCTION

Honda had in fact been extremely fortunate to take the 1983 500 title, their fortune arising primarily from the skills Spencer had been able to utilize to hang on to a clearly more powerful Yamaha. As the suspension on the OW70 had been sorted out, Spencer had found it almost impossible to hang on, and both he and Kanemoto made it clear to Japan that they needed a more powerful machine. Now it was Yamaha's turn to be fortunate as they made subtle but enormously important changes to the OW to give

The sophomore on his way to the graduation ceremony. In 1984, Lawson accepted the responsibility for the return of Yamaha's pride and brought back the 500 crown with a win here in Austria and in three other GPs.

Biggest Bang per Buck

birth to the OW76, whilst Honda developed a radical new machine that turned out to be a disastrous miscalculation.

The most important change that took place on the OW was the switch of induction regulation from a disc valve to reed valve. Reeds were of course nothing new for Yamaha, the company having been responsible for the resurrection of interest in the design in the early 1970s. It had been Honda though who re-thought how reeds might best be applied to racing machines, where top-end and mid-range power were equally important. Reeds had traditionally been mounted in the inlet tract passing through to an inlet port on a conventional piston-ported engine. The disc valve had been adopted in the search for the 'holy grail' that would provide total scavenging of the exhaust gases. The wall of the cylinder previously occupied by the inlet port could now be used for auxiliary transfer ports, all helping clean out exhaust gases. Traditional reed valve design wouldn't allow this.

There was however a price to pay, as the disc valve engine had initially been a poor starter and had a sharp transition between weak and strong power delivery. Both effects were caused by the earlier inlet timing used on disc valve engines to obtain more complete crankcase filling and consequently higher power delivery in the power band. At lower engine speeds, the carburettor is relying on the vacuum caused by the ascending piston to atomize and draw charge into the crankcase. If the inlet tract starts to open just after the piston has passed BDC as happens with early inlet timing, the piston is moving slowly, the vacuum is weak and the carburation suffers. The result was the legendary light switch.

In contrast, the reed valve engine would carburate better throughout the engine range as it was sucked open by the crankcase vacuum. It still had the disadvantage of restricting the flow at high engine speeds, but the large six-petal reed cages were far less restrictive than their predecessors. Mounted on the crankcases, the rear of the cylinder could still be used for extra transfer ports to assist scavenging. The better carburation would translate into

The naked 1984 OW76, the first of the V-4 reed valve Deltabox-framed 500s, looks like a raw prototype with its uneven welding and ill-fitting petrol tank. But it worked and set Yamaha on course for six championship wins in nine years.

smoother power delivery during the transition into the power band as the resonances in the exhaust pipes began to do their stuff. In addition, the exhaust seemed to be able to open the reed for a few degrees of crankshaft rotation just after BDC and draw in charge directly from the carburettor before there was sufficient vacuum under the rising piston to achieve this. Nobody was quite sure what was going on in the engine but it seemed to work. Disc valves were out and reed valves were in and they were to stay in for good.

The rest of the engine changes for the OW76 resulted from the move to reed valves. Both 34mm and 35mm flat-slide carburettors were used, the direct exposure of the carburettors to the pressure changes in the engine via the reeds causing the start of development of increasingly sophisticated instruments. The exhaust pipes were of a dramatically different design, with far longer gentler taper of the header pipe and dramatically sharp tapers to the centre section of the pipe. The overall length increased so much that the pipes on the two forward cylinders had to cross under the engine in order to fit within the length of the bike.

In comparison to the engine changes, only minor enhancements were made to the chassis. The spars of the twin-spar frame became wider and the small vestigial tubes holding the engine grew smaller. A cross-member of the frame arched up over the swing-arm axle, with the top end of the near vertical rear damper bolted to it. Some experimentation had been done on the length of the swing-arm during the winter, but the test had been inconclusive. The rear suspension geometry was very similar to that of the OW70, but the complete unit had been lowered in the struggle to keep the centre of gravity low. The total machine weighed in at 260lb (118kg) indicating the extensive use of trick metal. There was no question of replicas of this machine going on sale; it was literally almost priceless.

Typically it needed some early season work to get it sorted out. Kel Carruthers was, as always, responsible for this:

> At the beginning of the year only Eddie had the reed valve machine. Ferrari [Virginio Ferrari, Lawson's team mate] had to make do with the disc valve to begin with. The reed valve engine had round-slide carburettors and we went to the Match races (Transatlantic Trophy in the UK at Easter) and there were no Yamaha engineers there. The last thing they had told me was not to run the reed valve bikes. Eddie and Kenny had to ride the disc valve engines. The only time I ever can remember that Kenny was the number two rider. He got what Eddie didn't want.
>
> We decided it was a good opportunity to do some testing on the reed valve engines. They didn't run well with the standard carbs. Still in the truck I had some flat-side carburettors we'd been using on the OW70 the year before. To make it fit I had to cut a bit off the back of the carb and we put it on and it was a lot lot better. So then I had to call the factory and tell them I'd been a bad boy and run the reed valve engines with flat-side carburettors and it was like 100 per cent better. Then they sent all the 1983 carburettors for 1984.

Lawson and the OW76 ran away with the 1984 500 crown. Lawson was a good rider, getting better, the OW76 a good bike, getting better, and world champion Spencer was a superb rider at his peak, with a lousy bike with a lot of development to do. Honda built Spencer a V-4 reed valve 500, with very similar engine design to the Yamaha, with the exception of the single crankshaft rather that the double contra-rotating crankshafts of the OW. Honda had convinced themselves that friction within the engine was public enemy number one and it totally dominated their design philosophy for the rest of the decade. It has been argued that this

165

obsession with minimal engine friction has led them to become blind to problems caused by the engine designs they developed as a consequence. However, in 1984 the engine was not the problem, but the chassis. The places for the exhaust pipes and fuel tank were reversed, fuel stored in a tank slung under the engine, the pipes under a dummy tank above the engine. This lowered the centre of gravity all right, but caused the bike to become more difficult to 'flip' from side to side due to the rolling moment of inertia actually increasing. Honda had also started using carbon fibre on their machines and during practice for the first GP in South Africa, the carbon rear wheel broke up, spitting Spencer into the straw bales and resulting in some cracked bones in his foot. In fact Spencer was to crash several times during 1984, each time causing small bone fracture or breakages.

After four GPs struggling to sort out the NSR500 and dreaming of the good old days with the triple the year before, the NS500 was wheeled out for the German GP and ridden to victory. By then it was all too late for Spencer, with Lawson having an unassailable lead, and despite tyre trouble at several GPs causing him to slip to sub-rostrum places, he ended the season a clear thirty-one points ahead of Randy Mamola who had used the works NS500 to harry Lawson the whole season. After another late season crash, Spencer ended up fourth in the final world championships. It was the machine that had let him down as evidenced by the five wins and a second of the six races he finished. For 1984, Spencer was down, but definitely not out.

The 1984 OW76 had set a design template, a skeleton specification, featuring V-4 reed valve engine in a twin-spar chassis that was to last through to the present day. Since 1984, improvements made to the machines have centred around individual components that have been refined to function more effectively. Deeper analysis of the performance of the machine has become possible through the application of microelectronics and computer technology. Measurement is knowledge and the knowledge has enabled far better tuning of the racing motorcycle package, no longer totally reliant on the rider's analytical skills. The collective small advances have resulted in a significant lowering in lap times. One of the few circuits largely unchanged since 1984 is Mugello in Italy, where Takazumi Katayama set a lap record of 2 minutes 3.69 seconds in 1982 on a Honda NS500. Ten years later Mick Doohan on a Honda NSR 500 managed a lap of 1 minute 53.83 seconds. Both major revolutionary design changes and small evolutionary changes get the job done.

OW81

For the transition from OW76 to OW81, the minor changes involved a re-design of the primary drive, two extra boost exhaust ports not controlled by the guillotine powervalve, an extra exhaust box to provide variable volume exhaust pipes, the use of a full flange to hold the engine to the spar frame and 16 inch front and 17 inch rear wheel with Michelin Radials on both. Most interesting of these were the extra exhaust boxes, an idea copied from Honda who had been using these extra Automatic Torque Amplification Control (ATAC) chambers since 1982 to boost power at low engine speeds. Only the upper pipes of the OW81 were fitted with the boxes, but they were not found to offer significant improvements.

Honda, meanwhile, had junked the unconventional chassis design of the NSR500 to build a spar framed machine with everything back to its normal place. The engine turned out to be a touch stronger than the OW's, with little to choose as far as handling was concerned. The points totals at

Biggest Bang per Buck

By 1985, the OW81 essentially looked the same as the YZR500 Rainey rode to the 500 title in 1992. The design was proven – let evolution begin.

the end of the season suggest a close battle, but Lawson only managed to beat Spencer at the first GP at South Africa and later at Yugoslavia. A shredded rear tyre at Hockenheim and a fluffed start in France left Lawson with a couple of fourth places to Spencer's second and first respectively, and the crown was Freddie's.

There seems to have been remarkably little panic back in Iwata after the loss of the title. During the winter of 1985/86, Yamaha decided to do a little engine work to give slightly more top-end at the expense of a slight drop in power as the power band was entered. From 9,000rpm to 10,500rpm, there was strong power, with yet another kick of power through to 11,700rpm, after which it tailed off. The fifth transfer port at the rear of the cylinder wall opposite the exhaust port was widened and given a central bridge to support the piston ring. A little bit of pipework had also contributed to

This spy shot of Mike Baldwin's 1986 YZR500 clearly shows the gears on the end of the twin crankshafts.

167

Biggest Bang per Buck

the change in power delivery. The chassis had some scarcely discernible work on the rear suspension geometry easing the loads being passed to the Deltabox frame.

Gradually Yamaha had also been changing their policy for the supply of bikes to GP teams. Since their entry in the GP 500 class, a maximum of two teams received the complete OW each year. In 1985, Christian Sarron riding for Gaulois Sonauto, the French importer, had also received an OW81, with which he claimed the third place in the final standings behind Lawson. For 1986, a third team would be receiving Yamaha's latest weapon. Not only was there Agostini's Marlboro Yamaha and Gaulois Sonauto, but also a new team sponsored by Lucky Strike and managed by none other than Kenny Roberts.

> After the support we had been receiving from Yamaha for the 500s in 1983, we were really disappointed by the lack of support we got for our 250 team in 1984. We realized that Yamaha weren't going to do the work for us. At that time they weren't interested in 250s, so if you had a 250 you were on your own. They'd promise you help, but never give it you. So we had to drastically increase the budget to enable us to do it ourselves, but Marlboro didn't want to do it. I couldn't leave Yamaha as I had a contract with them. So there was nothing I could do for 1985. By 1986 we had the deal with Lucky Strike for the 500s.

YZR500

1986–9

As if realizing that a total of six riders on factory specials no longer justified the 'OW' epithet, the 500 was identified simply as the YZR500.

Honda in the meantime were in big trouble as Spencer was having serious problems with a virus during the winter and the development of tendonitis in his right arm. In addition the newly designed NSR500 seemed to have been another mistake with totally unpredictable steering. The problem

Christian Sarron was the only non-American rider to win a 500 race in thirty-eight GPs between 1982 and 1986. At a wet Hockenheim in 1985, he managed to stay 'wheels-up' and lead Freddie Spencer over the line.

A pumped-up Eddie Lawson realizes he won't convince the Dutch marshals to let him rejoin the 1986 TT after his first lap tumble. Fortunately he took the 500 crown anyway.

was later diagnosed as being caused by the direction of crank rotation causing the front wheel to lift at high revs. Wayne Gardner was set to fight six YZR500s on a camel of a machine designed for Freddie Spencer. It's no surprise that Lawson regained his crown in 1986, but not without some problems. Lawson was not happy with the change in power band and lack of chassis design as the Spanish GP kicked off the season. In addition, the carburation was proving difficult to set up correctly, and it cost him the win in Spain. After winning the next Italian GP, new carburettors helped him win at Hockenheim, although he still claimed the bike was difficult to start. New reed valves were available for Austria, another win, and by the next round at Yugoslavia, Lawson was declaring the bike to be perfect. By mid-season it was all over bar the shouting, Gardner managing to hang on to second place in the end-of-season results ahead of the other four YZR500s.

The 1987 YZR500 teams were as unchanged as unfortunately the bike itself turned out to be. Lawson and McElnea at Marlboro, Mamola and Mike Baldwin at Lucky Strike and Christian Sarron at Gaulois received machines with minimal changes from the 1986 machines. The chassis had been strengthened a little and there had been some experimentation with longer con-rods, compensated by spacers under the cylinders to maintain the same stroke. In contrast, the NSR Honda had been significantly adapted to match Gardner's riding style, now that he was the unchallenged number one rider. Kel Carruthers and Mike Sinclair did their best to construct competitive machines for Lawson and Mamola respectively, using cylinders and con-rods from the previous season, but it was to be Gardner's year. The internecine battle between Lawson and Mamola, Marlboro and Lucky Strike, Agostini and Roberts, Michelin and Dunlop, went to Mamola. This

Biggest Bang per Buck

One of the less fraught moments in the relationship between team manager Agostini and the quixotic Lawson. The 1986 title is theirs, the sponsors are happy and if you look at the T-shirt close enough, you'll see he rode a Yamaha.

was due more to good fortune than better bikes or riding skills, although the Dunlop rain tyres seemed significantly better in the three wet races of the year.

Bright spot of the 1987 season for Yamaha were the performances of Kevin Magee during his two European GPs at Assen and Jarama. A second fastest practice time was disappointingly only turned into tenth in the multi-start wet Dutch GP. At the Portuguese GP at Jarama, his fifth practice time became a third place in the race just behind teammate Randy Mamola. Once sponsorship problems were sorted out, there was a place for Kevin Magee for the whole 1988 season in the Lucky Strike team. Despite his second place in the final championship table, Randy was out of the team, replaced by the man Kenny Roberts had been waiting to sign up for the 500 class since their 1984 season running TZ250s, Wayne Rainey.

> After the year in Europe with the 250s, I returned to the US and raced a 250 and 500 for Bob Maclean. I did some dirt-tracking that year on a Harley. That was the most fun I had all year, wobbling round the race track. Kenny thought I was ready for Europe, but I needed some time to prepare myself mentally. For 1986 and 1987 I rode superbikes for Honda.

By 1988, with the US superbike title under his belt, Rainey was ready for Europe. Fortunately Yamaha had woken from its year-long slumber and had a bike ready for him, Kevin Magee, Eddie Lawson, Didier de Radigues and Christian Sarron, that had undergone small but significant changes. The bike steered a little quicker, something Lawson was particularly pleased to discover

Biggest Bang per Buck

What's the fun of being a team manager if you can't take the bikes out for a test drive once in a while? Kenny Roberts helped out with tyre tests during practice for the 1987 Italian GP. He was still very fast.

as it had been one of his major complaints about the 1987 bike. The most visible change to the engine was a widening by 10 degrees of the cylinder V. At 70 degrees, there was more room available for the carburettors and inlet tract. The inlet-side changes were complemented by a return to the cylindrical powervalve of old. Since the end of the 1970s, throttle control had been the key to successful riding of the 500s. Back-to-back tests between guillotine and cylindrical powervalves had shown the latter to provide more accurate throttle response. The 250s with their less than total requirement for rear-wheel steering retained the compacter guillotine valves.

The YZRs and the machines of other factories had been suffering for the last couple of seasons from the inability to fit the optimum exhaust pipe design. The exhausts of the rear two cylinders passed under the rider's legs to exit from the tail of the seat unit. Up until 1988, the pipes from the front cylinders passed back under the engine, hugging the swing arm. The maximum diameter of the middle section of the pipe was limited by ground clearance and the length of the section was limited by the fact that it had to converge again before reaching the swing-arm. Yamaha solved the problem by designing a swing arm whose left-hand side was a conventional braced spar, but whose right-hand side curved up and over two dimensionally correct exhaust pipes. The first of the 'gull-wing' swing-arms had come into existence. The other manufacturers soon followed suit.

Thus far, all YZR500 riders were equal, but the Marlboro and Gaulois teams had the advantage of Michelin tyres who at the time

171

The 500 GP of the decade? The Paul Ricard circuit 1988 with Gardner (1) and Schwantz (34) filling the Yamaha sandwich of Sarron (7) and Rainey (17). The injured Lawson moved through the field to take the win after Gardner's machine blew up with two corners to the flag.

seemed to offer better compounds to their riders. Also Lawson and Sarron had been campaigning YZRs for several years, giving them a clear advantage over the three other rookies. The Marlboro team considered themselves to be the factory team whilst Lucky Strike seemed to have an arm's length relationship with Yamaha. This meant that they got less direct support from the factory and felt free to make up this deficiency by actively developing the bikes themselves. It put a lot of strain on the riders, in Europe for a full first season, learning the circuits, learning to ride state-of-the-art 500s and develop them further. Rainey recalls the absence of overt pressure on the riders

> It was a learning year for me and I wasn't real concerned that we had to beat everybody. Nobody put pressure on me to perform. It was just step, step, step. We had some good races, got on the podium about half of the races and won at Donnington.

Magee was the first to bring the Lucky Strike team a GP win, when he played the backmarker game with Eddie Lawson to his advantage and won the Spanish GP at Jarama. The Lucky Strike team seemed able to beat both other Yamahas and Gardner's Honda, but for problems with Dunlop tyres. Evidence of the development work being done on the bikes could be seen in the form of welds on the swing arm where it had been shortened, the mid-season appearance of upside-down Öhlin OUT forks and the addition of extra radiators to bring the operating temperature of the engine down to around 65°C. Yet another development was visible at Donnington Park, scene of Rainey's first GP win.

> I fitted AP carbon discs for the first time at Donnington. You had to be real careful to remember to keep them hot. I almost crashed on the warm-up lap as they were too cold. Almost took Schwantz out at the Old Hairpin, so I braked hard for the rest of the lap, and they worked fine after that.

At the end of the season Rainey finished third behind Gardner and the best rider of

Biggest Bang per Buck

the late 1980s, Eddie Lawson. This was Kel Carruthers' sixth and last 500 title, three apiece with Eddie Lawson and Kenny Roberts. His role as mentor, team manager, conduit to Japan, chief engineer, and sage, was the catalyst that translated the enormous skills of both riders into world championship victories. Kel would be happy to remember 1988 as his last active year working on the YZR500s, as 1989 turned into a huge disappointment for the Agostini Marlboro team of Freddie Spencer and Niall Mackenzie. Yamaha sought to honour Kel with the job of global race co-ordinator for 1990, but it was not the right sort of work for a man happiest with his hands deep in the guts of a racing motorcycle. Kel broke the link with Yamaha after 1990 and went freelance marking the end of a twenty-year long relationship between the two. No single individual can be given all the credit for the great achievements of the factory within the world of motorcycle racing during this period. But Kel Carruthers can be given much of it.

Little seemed to have changed on the YZRs for 1989 as they lined up in Suzuka for the first GP of the year. In fact several of the experiments that had been cautiously applied during 1988 were developed during the winter and incorporated into the new model. Both the Marlboro and Lucky Strike

The metamorphosis of the early V-4 OW series into the 1994 YZR500 was essentially complete by the 1989 season. With 'gull-arm' rear swing-arms and Öhlins upside-down forks, they brought Rainey within an ace of the championship title. These are Sarron's machines.

Biggest Bang per Buck

team had been experimenting with extra radiators during the 1988 season, the new thinking in the two-stroke world being that the engine should run around 65°C rather than the 75°C that had been the practice at the time. The theory was that the lower engine temperature would lower the temperature of the charge in the crankcase, providing higher charge density. Flat-crown pistons were also fitted for the first time in the 'four-port' cylinders still used on the YZR. The domed pistons previously used would have compromised (however slightly) the design goal of achieving a slowly rising column of fresh charge in the centre of the cylinder effectively scavenging exhaust gases. The usual juggling of transfer port flow angles suggested attempts by the factory to direct the charge further to the rear of the cylinder, again to keep the charge from prematurely disappearing out through the gaping exhaust port. Both changes helped to some, extent but didn't provide an enormous increase in engine performance.

On the chassis side, there had been a major revelation at the end of the 1988 season. Rainey explains:

> The 1988 season we'd had real problems with tyres. We'd tried so many different compounds and none of them seemed to work well. After the last GP in Brazil, we stayed on and Kenny and I tested about 300 different tyres, without finding the right combination. Then we bolted some 17 inch rear wheels on and suddenly we were there. It was a quantum jump in performance.

This discovery, coupled to the hope that Dunlop would come up with a better product than Michelin, resulted in the team continuing into 1989 with the Japanese/British company. In retrospect it was a wrong decision, for although the tyres were better than the previous year, there were still problems with side-wall grip that were not sorted during the season. Despite this, it looked like being Rainey's year as Lawson, now on a Rothman's Honda, struggled to turn the

Behind every great rider, there's a great team. Five years later everyone on this photo still worked for Kenny. From left, Skip Martin, Mike Sinclair, Warren Willing, Trevor Tilbury, Bernard Ansiau, H.Gregory, KR and Thierry Gerin.

Biggest Bang per Buck

The third place at Donnington in 1989 was not an especially good result for Rainey, but he still left the UK with a six-point lead over Lawson. The fall in Sweden a week later cost him the 1989 title.

bike into a Yamaha and the Marlboro Yamaha challenge of Spencer and Niall MacKenzie never materialized. Rainey headed the series right through to the Swedish GP, when he crashed out while challenging Lawson for the lead. It was too late in the season to claw back the lost points and Lawson took his last 500 title by seventeen points.

The 1990s

The single most significant change in the evolution of the YZR as it entered its seventh year as a reed valve V-4 was the change to a six-port engine. Two extra transfer ports were squeezed into the walls of the cylinder, causing the transfers closest to the exhaust to start creeping underneath the small auxiliary exhaust booster ports. Bud Aksland doing the dyno work for both the 250 and 500 team, believes this change added a good 10bhp to the output of the engine, now up around the 165bhp range. The new engine slotted into a chassis running on Michelin tyres, resulted in the best package for 1990. At the time, Rainey described the changes brought by the Michelins as follows:

> The Michelins give more side grip when you accelerate, there is no doubt about that, but there is less warning of a slide. With the Dunlops, you could slide all day with a good chance of being able to control the slides. That was great, but very frustrating when you were chasing someone on Michelins who was able to get on the power earlier in the corner because he was getting more grip. With the Michelin you can put in a better lap time, but you had better concentrate. You can't afford to relax for a second because when they do let go, they do it in a real hurry.

It was an extraordinarily effective season of racing by Rainey, winning seven GPs, on the rostrum at all GPs except Hungary when brake failure forced retirement. With John Kocinski taking the 250 title, it was an exceptionally good year for the Marlboro

Biggest Bang per Buck

Gotcha!! Rainey wins the 1990 Czechoslovakian GP and his first 500cc crown. Later that day ...

... Rainey's caught admiring the number still adorning Eddie Lawson's machine. He'd have to wait another few months before he could use it on his own bike.

Black Beauty. Steve Blackburn and Thierry Gerin put the finishing touches to Kocinski's 1991 YZR500. Looks like it's been a long night. The bike was bristling with sensors to measure everything, most conspicuous being the travel sensors on the front fork.

team of Kenny Roberts. Despite the clear superiority of the package, there were updates through the season, including a new chassis with the engine 15mm higher and further forward than the previous version. Also the swing arm was mounted in eccentric bearings allowing its height to be changed by up to 4mm. Aware of the 10 to 18lb (5 to 8kg) weight penalty of the YZR with respect to the Suzukis and Hondas, titanium exhaust pipes were used at the Czechoslovakian GP.

It looked like there would be more of the same in 1991, with the sole complicating factor being a change back to Dunlop tyres. Late in the 1990 season, Michelin HQ had been crunching some numbers to analyse their financial performance and decided that it was not good enough to continue to support the GP racing scene. Somewhat prematurely, the company announced its withdrawal from the sport. A few weeks later it turned out that the numbers were not so bad after all and a limited return to racing was announced for 1991. The Roberts team chose to switch back to Dunlop with the promise of a fully supported programme. In the end, both manufacturers ended up having problems with their products, Rainey having to change a disintegrating rear tyre at Misano and consequently only managing a ninth place. Rainey took the 1991 title after trailing Mick Doohan, on the Rothmans Honda, right through to the Dutch TT where Doohan took a tumble and lost the points initiative. Brilliant but inconsistent rides by Kevin Schwantz brought him into a close third place in the final points table.

There was as always little new equipment coming out of Japan during the season until a new chassis arrived for the Donnington Park GP. The chassis was intended to complement the work Öhlins had been doing on their rear suspension units. This chassis was run using a conventional rear suspension unit, but at the San Marino GP two weeks later the same chassis was used with Öhlins Computerized Electronic Suspension system (CES). This enabled the damping to be adjusted via a computer that could be plugged in to the unit without it being removed from the bike. Damping adjustments could be made in seconds, a valuable asset when setting up the bike.

The successes of the Marlboro Yamaha race team unfortunately did not reflect the good health of 500cc class racing. From the mid-1980s most countries had stopped running national and international 500 class races. The RG500 Suzukis had been the last reasonably priced racer of the class. The RS500 Honda was too expensive to be sold in large numbers to riders other than those with GP-level budgets on which to go racing. In the GPs, there was an ever-widening gap between the factory riders and the RS500 riders, who seldom made the top ten positions and missed out on any significant prize money to ease their acute cash flow problems. The rich were getting richer and the poor poorer. Half-hearted plans to make year-old factory machines available for leasing were not successful, the $400,000 price tag frightening off all but the richest of non-works riders. By 1991, the situation had reached crisis point, with 500 class grids shrinking to less than twenty riders, instead of the thirty plus of a decade earlier. One of the motorcycle's greatest attractions, close racing throughout the field, was missing, and the spectators were turning to the 250 class for action and interest. The 500 class was turning into a support class for the real racing.

Many of those working within the 500 class were unhappy with these developments, but few were in a position to do anything about it. Kenny Roberts was one of those few and he exercised his considerable influence with Yamaha to convince them of the need for corrective action. Taking the Formula 1 car industry as an example, he

177

brokered a deal between Yamaha and European chassis developers committing Yamaha to supply engines and spares to be slotted into chassis built in Europe.

The price was set by Yamaha at £75,000 for the complete bike with spares for half a season's racing. A second spares kit could be bought for another £7,500 during the season. Fifty thousand pounds went to Yamaha for the engine, carburettors and electrics, and £5,000 for the Öhlin forks and rear shock absorber. The rest of the machine had to be bought in or constructed by the company selling the bike. Some constructors were approached who turned down the proposal as financially unviable for them, but Steve Harris in the UK and Serge Rosset of ROC took up the deal. Rosset also helped make the deal more profitable for ROC by including a service package costing an extra £25,000 which would give the rider the right to support with the set-up of the bikes at each of the GPs. Rosset had been looking after the Sonauto YZRs for some years and had the data available for each of the circuits.

The Harris-Yamaha was a replica of the 1990 YZR that had brought Rainey his first world title. The bike had been mothballed at the Marlboro HQ and was dusted off and delivered to Steve Harris's workshop. In contrast, the ROC YZRs were replicas of the YZRs that Christian Sarron and Jean-Phillipe Ruggia had run in 1991. This led to the ROC Yamahas having adjustable engine and rear swing-arm position, something not possible on the Harris-Yamahas. In total eight ROC and six Harris machines were assembled, their owners dreaming of mixing it with the £2,000,000 factory machines.

Unfortunately, it was to remain a dream, the best non-factory YZR placing being Paul Goddard's fifth place at the Donnington GP after several of the top riders retired or crashed. There was some doubt about the geometry of the Harris chassis, and the ROCs definitely seemed to perform better. All the privateer riders took a long time to get used to the bike's 150bhp, as witnessed by a narrowing of the lap time gap between the works and non-works YZRs as the season progressed. The cost of racing these machines was offset by the International Racing Teams Association (IRTA) re-alignment of start pay and prize money. Despite not doing a whole lot better in positions through out the year the privateers managed to survive without breaking their shoestring budgets.

One organization not working to a shoestring budget was Honda. 'Whatever it takes' was their motto and once again it was demonstrated at the first GP of the year at Suzuka. Motorcycle racing had always stimulated more sensory organs than other sports, the ears coming a close second to the excitement generated by the images picked up by the eyes. Veterans of the sport talk with glistening eyes of the wonderful sounds of the four-stroke/two-stroke battles of the 1960s. The demise of MV had robbed the grids of the throaty roar of the four-stroke, but the differing engine configurations run by Suzuki, Honda and Yamaha had maintained their aural identity through to the middle 1980s. With the universal adoption of the V-4 configuration, this aural identity had been lost. Until practice for the first GP of the year started. A strange drone-like burble could be heard from the Honda garage as the 1992 NSR was being warmed up. The first of the 'big-bang' engines was preparing to humiliate its competitors.

The NSR had been designed so that both pairs of cylinders fired almost simultaneously, probably within about 70 degrees of crankshaft revolution. This resulted in no higher power levels than before, but torque delivery, the engine's pushing power, was smoother, maintaining a high level for a longer fraction of the crankshaft revolution. Although seeming to be slow to the rider, the

Biggest Bang per Buck

The privateers' dreams of mixing it with the factory lads turned sour as the ROC- and Harris-Yamahas proved tricky to set-up and run. Peter Goddard had the most success with his ROC Yamaha.

bike utilized the power it was generating more effectively. It demonstrated an ability to hook up the tyres on the exit from corners which the riders of the old 500 screamers with their 180 degree firing interval, could only marvel at from an ever-increasing distance. Mick Doohan won five of the first seven GPs and came second in the other two. He got pole position at six of the seven GPs. It was going to be Honda and Doohan's year at last.

The competition were rushing to produce their own versions of the big bang engine, Suzuki getting there for GP number seven, followed by Cagiva at the next GP and Yamaha, as always slow with mid-season development, another GP later. The re-

design was not insignificant as not only were different crankshafts built, but crankcases, primary transmission and clutch also needed to be strengthened to withstand the sustained torque levels being produced. Indications of the rush that had been put into getting the engine ready manifested themselves as crank failures on the Suzuki and primary transmission failure at the British GP for two of the four big bang Yamaha engines.

The new engines were the big news of the year, and all other changes to the Marlboro YZR were minor in detail. New fairings were visible at the second GP in Australia. A revised chassis was built for the team by ROC, with the factory's blessing, and

179

Biggest Bang per Buck

Everything seemed to be going wrong for the Marlboro team in 1992 until the European GP at Barcelona. Honda's new engine had surprised everyone and only here at the tight Catalonya circuit could Rainey hang on to Doohan and squeeze past close to the end. One of his best races ever.

appeared at Barcelona for the GP of Europe. At Hungary, Rainey joined the ranks of riders running ignition interrupt switches on their gear pedals to enable them to make clutchless full throttle upshifts. New Öhlins forks arrived at the French GP, but were not effective until the subsequent GP at Donnington. All the improvements made a good bike better, but they would not have brought Yamaha the world title they and Rainey achieved in 1992. Mick Doohan was unable to complete his season competitively after breaking a leg at Assen. With the best rider/bike package of 1992 sidelined, Rainey was able to turn the sixty-five point mid-season deficit into a four-point lead after the South African GP, ending the 1992 season. This was only the second time Yamaha had achieved a hat-trick of world titles, a feat previously managed by Rainey's team manager Kenny Roberts.

Post-season tyre tests seemed to confirm that the strengthened engine had the reliability it would need for the GPs of 1993. By about 300 miles (500km), both cranks and pistons were approaching failure, but that would easily allow them to last a GP. It was decided therefore to ask the factory to concentrate on the chassis, although stronger aluminium engine cases were used in place of the magnesium ones previously fitted. The new chassis that Yamaha made available to the Roberts team of Rainey and new boy Luca Cadalora was extruded rather than of the welded-up Deltabox design they had been using since 1984. As before the engine mountings enabled the engine position to be changed slightly to facilitate setting the bikes up at each circuit. The extruded design increased chassis stiffness still further than in the past and it was hoped that with the best tyres Dunlop could provide, the YZR500 for 1993 would be more than a match for the NSR500 and RGV500.

Alarm bells started ringing when pre-season tests at Phillip Island in Australia had Rainey and Cadalora complaining bitterly about evil handling bikes. A few weeks later at Laguna Seca, testing showed some improvements, but there was a nagging suspicion that they had taken a wrong turn in the change of chassis. It turned out to be too stiff, removing the element of frame twist from the overall suspension function of the YZR. Cadalora in particular had difficulty with the new bike, leading to the inevitable criticism that 250 riders could not ride 500s. Within a couple of GPs, it was clear that the best 500 rider of his generation was also unhappy with the YZR; Team Roberts were in trouble. Despite the problems, Rainey managed good places at the first GPs in the Far East. A new frame available for the Japanese GP seemed to be a good deal better and he took the win at Suzuka. The bad news was that any further changes required by the team could not be made by Yamaha as the small resource-bound racing department could not provide the responsiveness that would be needed for the tight fortnightly European GP schedule.

A second place at the Spanish GP at Jérez seemed on paper to be fine, but if Rainey wasn't winning Kevin Schwantz was. The lack of consistency that had kept Schwantz from the World Title for five years on the GP trail had finally been pinned down by the Suzuki team. At the following circuits of Salzbürgring and the Hockenheim, Schwantz was up at the front going for the win, whilst Rainey could only struggle home third and fifth. The new points system for 1993 magnified the gap in performance and Rainey went to the mid-season Dutch GP with a fourteen-point deficit. The third chassis of the year was available for Rainey at Assen, this time manufactured at extremely short notice by Serge Rosset's ROC company and modelled closely on the 1991 YZR chassis used by ROC for their YZR production racers. Despite a time during free practice a full second faster than the full factory bike, Rainey stuck to the factory machine for the race and finished fifth for his pains to Schwantz's win. It was clear to the 120,000 spectators at the Dutch TT that there were serious handling problems with the machine with the back-end breaking away viciously at the exit of most of the slow corners. Rainey's title chances seemed all but dead, nothing short of a miracle being needed to bring him back into contention.

The miracle took place a few weeks later in the British GP at Donnington, in the form of a late-breaking Mick Doohan scuttling both Schwantz and team-mate Alex Barros. It was to be a Marlboro Team Roberts 1–2, although Rainey, battered from a spill during practice, could not hold off Luca Cadalora charging to the first 500 GP win of his career. Schwantz left the UK with no extra points and what turned out to be a broken left hand. Now it was his turn to struggle home in fifth place to Rainey's win at the Czechoslovakian GP, and lose the points lead as a result. On to Misano for the Italian GP and the dream became a nightmare. Looking set to take the GP and building up a lead at the front of the pack, Rainey braked a touch too deep on a high-speed right-hander. The bike was disturbed by a slight break in the tarmac at that point and lurched sideways, the front-end tucking under and pitching Rainey down the road in what looked like a nasty, but none too serious routine crash. As he lay at the side of the track, clearly conscious but not attempting to sit or stand, it became increasingly clear

Biggest Bang per Buck

that this was no routine crash. Hours later it was announced that his back was broken and he would most likely never walk again.

It seemed incomprehensible that this could have happened to Rainey. If ever there had been a rider who had demonstrated that it was possible to ride the 170bhp machines fast but safely, it was Rainey. He had ridden whole seasons during the infamous 'high-side' years at the end of the 1980s without crashing, while others bit the dust with sickening frequency. As tyres became more evenly matched to the power characteristics of the engines driving them, the danger of high-sides receded as did the number of crashes in the 500 class. Rainey's accident illustrated the dangers of the sport to those who had lived through the years of spectacular high-side crashes and had become blasé about the seemingly harmless low-siders. That it should happen to anyone is a tragedy; that it should happen to Rainey was especially cruel.

Alt hough it was up to Team Roberts to challenge for the world title in Yamaha's name, the ROC and Harris-Yamahas were out in even greater numbers than the year before. Serge Rosset was offering ROC riders, with enough cash, a package deal involving total support for the team before, during and after each GP. It was the key to success for many of the privateers, the complexities

The 1993 Harris-Yamaha's most notable difference from its predecessor was the carbon fibre sub-frame. Again, the ROCs with their all-in service package gave the richer teams a better chance of collecting valuable championship points.

of setting up a 500cc GP racer requiring years of experience for mastery. Only John Reynolds and Sean Emmett among the Harris riders were able to get both bike and body coordinated to finish consistently in the points. Hopes for supplies of big-bang engines to private teams were not forthcoming, although Niall Mackenzie's third place at Donnington showed that the old engines were still competitive.

Yamaha's service to GP racing can be seen when the final championship positions for 1993 are examined. Twenty-eight of the forty riders obtaining points were riding Yamaha-powered racing motorcycles. The malaise that had seemed to threaten the future existence of the 500 class just a couple of years before was gone. It had been cured by a combination of an organizational revolution in the structure of the sport and Yamaha's commitment to keeping the Summer Sunday a day for racing. Its future seems secure, although the decreasing percentage of sponsorship money coming from the tobacco industry and the increasing restrictions on its access to television advertising has cast an ever lengthening shadow on Grand Prix racing. With the tight advertising budgets caused by the recession of the early 1990s, alternative sources of sponsorship revenue are not immediately obvious.

As we enter Yamaha's third decade of 500cc racing and the fourth decade of TD/TZ250, their commitment seems undiminished with a single cylinder TZ125 joining its two big brothers in the company's road racing portfolio. No other company in the history of motorcycling has shown this degree of long-term support for racing. These years have seen the two-stroke engine develop from a hopelessly inefficient utilitarian powerplant into the fearsomely powerful unit at the pinnacle of internal combustion engine design today. Yamaha has been in the vanguard of those determined to understand and harness the forces at work within the two-stroke engine. It has been a long hard struggle along the road to true enlightenment; the 1993 TZ250 and YZR500 are a testament to Yamaha's achievement so far.

Appendix
250CC MOTORBIKE SPECIFICATIONS

MODEL	TD1	TD1A	TD1B	TD1C	TD2	TD2B	TD3	TZ250A	TZ250B
	Parallel Twin	Parallel Twin	Parallel Twin	Parallel Twin	Parallel Twin	Parallel Twin	Parallel Twin	Parallel Twin	Parallel Twin
Year	1962	1963	1965	1967	1969	1971	1972	1973	1975
Model code	D6-001+	D6-060+	D6-120+	TDIC-000101	DS6-900101	DS6-900501	DS7-990101	DS7/430-991101	DS7/430-991471
Bore (mm)	56	56	56	56	56	56	54	54	54
Stroke (mm)	50	50	50	50	50	50	54	54	54
Capacity (cc)	247	247	247	247	247	247	247	247	247
Compression ratio	9.00	9.00	8.10	8.10	7.60	7.60	7.60	7.60	7.60
bhp	32	32	35	40	44	47	49	51	51
@ rpm	9,500	9,500	9,500	10,000	10,500	11,000	10,500	10,500	10,500
Ignition system	magneto	magneto	magneto	magneto	magneto	magneto	CDI	CDI	CDI
Ignition timing (mm)	1.7	2.1	2	2	2	2	2	2	2
Carburettor	Mikuni VM20H	Mikuni VM20H	Mikuni VM20H	Mikuni VM30SC	Mikuni VM30SC	Mikuni VM30SC	Mikuni VM34SC	Mikuni VM34SC	Mikuni VM34SC
Primary drive	3.25	3.25	3.25	3.65	3.7	3.7	3.22	3.35	3.35
Final drive	38/16	34/17	34/17	35/19	34/20	34/20	34/16	34/16	34/16
Box gearing sixth	–	–	–	–	–	–	0.81	0.81	0.81
Box gearing fifth	0.75	0.92	0.92	0.95	0.95	0.9	0.87	0.87	0.87
Box gearing fourth	0.96	1.04	1.04	1.05	1.05	1	0.96	0.96	0.96
Box gearing third	1.23	1.23	1.23	1.17	1.24	1.18	1.13	1.13	1.13
Box gearing second	1.66	1.58	1.58	1.53	1.53	1.47	1.42	1.42	1.42
Box gearing first	2.5	2.5	2.27	2	2	2	1.93	1.93	1.93
O/A top gear ratio	5.94	6	6	6.39	5.97	5.66	5.43	5.78	5.78
Fuel (l)		21	21	21	23	23	23	23	23
Box oil (cc)	1,100	1,500	1,500	1,500	1,500	1,500	1,500	1,500	1,500
Front wheel	2.50 x 18	2.50 x 18	2.50 x 18	2.75 x 18	2.75 x 18	2.75 x 18	2.75 x 18	2.75 x 18	2.75 x 18
Rear wheel	2.75 x 18	2.75 x 18	2.75 x 18	3.0 x 18	3.0 x 18	3.0 x 18	3.0 x 18	3.0 x 18	3.0 x 18
Front brake	tls drum	tls drum	tls drum	tls drum	dtls drum	dtls drum	dtls drum	dtls drum	dtls drum
Rear brake	sls drum	sls drum	sls drum	sls drum	sls drum	sls drum	sls drum	tls drum	tls drum
Front suspension	teles	teles	teles	teles	teles	teles	teles	teles	teles
Rear suspension	dual shock	dual shock	dual shock	dual shock	dual shock	dual shock	dual shock	dual shock	dual shock
Wheelbase (mm)	1,295	1,295	1,290	1,280	1,315	1,315	1,300	1,300	1,300
Width (mm)	610	737	610	540	510	510	510	510	510
Length (mm)	1,943	1,943	1,950	1,887	1,925	1,925	1,940	1,940	1,940
Rake	–	–	–	27.5°	27.5°	27.5°	27.5°	27.5°	27.5°
Trail (mm)	–	–	–	88	90	90	90	90	90
Dry weight (kg)	111	96	96	104	105	105	105	110	110

Appendix

MODEL	TZ250C	TZ250D	TZ250E	TZ250F	TZ250G	TZ250H	TZ250J	TZ250K	TZ250L
	Parallel Twin	Parallel Twin	Parallel Twin	Parallel Twin	Parallel Twin	Parallel Twin	Parallel Twin	Parallel Twin	Parallel Twin
Year	1976	1977	1978	1979	1980	1981	1982	1983	1984
Model code	DS7/ 430-993101	DS7/ 430-994101	DS7/ 430-995101	DS7/ 430-997101	4AI 000101	5F7- 000101	–	26J- 00101	49V- 00101
Bore (mm)	54	54	54	54	54	56	56	56	56
Stroke (mm)	54	54	54	54	54	50.7	50.7	50.7	50.7
Capacity	247	247	247	247	247	249	249	249	249
Compression ratio	7.85	7.80	7.80	7.80	7.60	7.60	7.60	7.80	7.80
bhp	52	53	53	53	55	57	57	59	60
@ rpm	10,500	10,500	10,500	10,500	11,500	11,000	11,000	11,000	11,000
Ignition system	CDI	CDI	CDI	CDI	CDI	CDI	CDI	CDI	CDI
Ignition timing (mm)	2	2	2	2	2	1.2	1.2	1.2	1.2
Carburettor	Mikuni VM34SC	Mikuni VM34SC	Mikuni VM34SC	Mikuni VM34SC	Mikuni VM34SC	Mikuni VM36	Mikuni VM36	Mikuni VM38	Mikuni VM38
Primary drive	3.35	3.35	3.35	3.35	3.35	3.15	3.15	3.15	3.15
Final drive	34/16	34/16	34/16	34/16	34/16	35/17	35/17	35/17	35/17
Box gearing sixth	0.81	0.81	0.81	0.81	0.81	0.91	0.91	0.91	0.91
Box gearing fifth	0.87	0.87	0.87	0.87	0.87	0.96	0.96	0.96	0.96
Box gearing fourth	0.96	0.96	0.96	0.96	0.96	1.08	1.08	1.08	1.08
Box gearing third	1.13	1.13	1.13	1.13	1.13	1.26	1.26	1.26	1.26
Box gearing second	1.42	1.42	1.42	1.42	1.42	1.56	1.56	1.56	1.56
Box gearing first	1.93	1.93	1.93	1.93	1.93	2.07	2.07	2.07	2.07
O/A top gear ratio	5.78	5.78	5.78	5.78	5.78	5.92	5.92	5.92	5.92
Fuel (l)	23	23	23	23	23	23.5	23.5	23.5	23.5
Box oil (cc)	–	–	–	–	–	–	–	–	–
Front wheel	2.75 x 18	2.75 x 18	2.75 x 18	2.75 x 18	3.0 x 18	3.0 x 18	3.0 x 18	3.25 x 18	3.25 x 18
Rear wheel	3.0 x 18	3.0 x 18	3.0 x 18	3.0 x 18	3.5 x 18	3.75 x 18	3.75 x 18	3.75 x 18	3.75 x 18
Front brake	disc	disc	disc	disc	disc	disc	disc	disc	disc
Rear brake	disc	disc	disc	disc	disc	disc	disc	disc	disc
Front suspension	teles	teles	teles	teles	teles	teles	teles	teles	teles
Rear suspension	monoshock	monoshock	monoshock	monoshock	monoshock	monoshock	monoshock	monoshock	monoshock
Wheelbase (mm)	1,315	1,315	1,315	1,320	1,320	1,320	1,320	1,320	1,320
Width (mm)	630	630	630	630	630	615	615	615	615
Length (mm)	1,930	1,930	1,930	1,935	1,935	1,950	1,950	1,950	1,950
Rake	27.5°	27.5°	27.5°	26°	26°	24.5°	24.5°	24.5°	–
Trail (mm)	74	74	74	97	92	87	87	87	–
Dry weight (kg)	118	118	118	107	107	106	106	103	106

Appendix

MODEL	TZ250N	TZ250S	TZ250T	TZ250U	TZ250W	TZ250A	TZ250B	TZ250D	TZ250E
	Parallel Twin	Parallel Twin	Parallel Twin	Parallel Twin	Parallel Twin	Parallel Twin	V-twin	V-twin	V-twin
Year	1985	1986	1987	1988	1989	1990	1991	1992	1993
Model code	59W-00101	IRK-00101	2KM-00101	3AK-00101	3LC-00101	3TC-00101	3YL-00101	4DP-002101	–
Bore (mm)	56	56	56	56	56	56	56	56	56
Stroke (mm)	50.7	50.7	50.7	50.7	50.7	50.7	50.7	50.7	50.7
Capacity	249	249	249	249	249	249	249	249	249
Compression ratio	8.1						8.3:1	8.3:1	
bhp									
@ rpm									
Ignition system	CDI	CDI	CDI	CDI	CDI	CDI	CDI	CDI	CDI
Ignition timing (mm)	–	–	–	–	–	–	variable	variable	variable
Carburettor	Mikuni VM38	Mikuni VM38	Mikuni TM38	Mikuni TM38	Mikuni TM38	Mikuni TM38	Mikuni TM38	Mikuni TM38	Mikuni TM38
Primary drive	3.15	3.15	3.15	2.6	2.6	2.6	2.6	2.6	2.6
Final drive	36/17	36/17	36/17	38/17	36/16	37/16	36/15	36/15	36/15
Box gearing sixth	0.91	0.91	0.91	0.91	0.91	0.91	0.91	0.91	0.91
Box gearing fifth	0.96	0.96	0.96	0.96	0.96	0.96	0.96	0.96	0.96
Box gearing fourth	1.08	1.08	1.08	1.08	1.08	1.08	1.08	1.08	1.08
Box gearing third	1.26(1.24)	1.26(1.24)	1.26(1.24)	1.26(1.24)	1.24	1.24	1.24	1.24	1.24
Box gearing second	1.56(1.48)	1.56(1.48)	1.56(1.48)	1.56(1.48)	1.48	1.48	1.48	1.48	1.48
Box gearing first	2.07(2.0)	2.07(2.0)	2.07(2.0)	2.07(2.0)	2.00	2.00	2.00	2.00	2.00
O/A top gear ratio	6.07	6.07	6.07	5.29	5.32	5.47	5.67	5.67	5.67
Fuel (l)	23.5	23.5	23.5	23.5	23.5	23.5	23.5	23.5	23.5
Box oil (cc)	500	500	500	500	500	500	500	500	500
Front wheel	3.0 x 17	3.25 x 17	3.25 x 17	3.5 x 17	3.5 x 17	3.5 x 17	3.75 x 17	3.75 x 17	3.75 x 17
Rear wheel	3.75 x 18	3.75 x 18	3.5 x 18	4.0 x 17	4.0 x 17	5.25 x 17	5.25 x 17	6.0 x 17	6.0 x 17
Front brake	dual disc	dual disc	dual disc	dual disc	dual disc	dual disc	dual disc	dual disc	
Rear brake	single disc	single disc	single disc	single disc	single disc	single disc	single disc	single disc	
Front suspension	teles	teles	teles	teles	teles	teles	UPSD	UPSD	UPSD
Rear suspension	monoshock	monoshock	monoshock	monoshock	monoshock	monoshock	monoshock	monoshock	
Wheelbase (mm)	1,325	1,325	1,325	1,336	1,335	1,312	1,328	1,328	1,328
Width (mm)	615	615	615	615	615	650	650	650	650
Length (mm)	1,945	1,945	1,955	1,936	1,935	1,920	1,942	1,942	1,942
Rake	26.5°	26.5°	23.5°	23.5°	23.5°	23°	22.5°	22.5°	22.5°
Trail (mm)	90	90	83	83	83	84	81.5	81.5	81.5
Dry weight (kg)	103	103	103	103	99	98	99	99	99

Appendix

350CC MOTORBIKE SPECIFICATIONS

MODEL	TR2	TR2B	TR3	TZ350	TZ350B	TZ350C	TZ350D	TZ350E	TZ350F	TZ350G
Year	1969	1971	1972	1973	1975	1976	1977	1978	1979	1980
Model code	R3-900101	R3-900501	R5-990101	R5/381-990101	R5/383-991001	R5/383-992001	R5/383-993001	R5/383-99401	R5/383-997001	R5/383-997504
Bore (mm)	61	61	64	64	64	64	64	64	64	64
Stroke (mm)	59.6	59.6	54	54	54	54	54	54	54	54
Capacity (cc)	348	348	347	347	347	347	347	347	347	347
Compression ratio	6.50	6.50	7.04	7.04	7.40	7.40	7.50	7.50	6.90	6.90
bhp	54	56	58	60	60	62	64	64	72	72
@ rpm	9,500	10,000	9,500	9,500	9,500	10,000	10,500	10,500	11,000	11,000
Ignition system	magneto	magneto	CDI	CDI	CDI	CDI	CDI	CDI	CDI	CDI
Ignition timing (mm)	2	2	2	2	2	2	2	2	2	2
Carburettor	Mikuni VM34SC	Mikuni VM34SC	Mikuni VM34	Mikuni VN34SC	Mikuni VM34SC	Mikuni VM34SC	Mikuni VM34SC	Mikuni VM34SC	Mikuni VM38SS	Mikuni VM38SS
Primary drive	2.7	2.7	2.96	2.96	2.96	2.96	2.96	2.96	2.96	2.96
Final drive	35/17	35/17	35/17	34/16	34/16	34/16	34/16	34/16	34/16	34/16
Box gearing sixth	–	–	0.81	0.81	0.81	0.81	0.81	0.81	0.81	0.81
Box gearing fifth	0.82	0.82	0.87	0.87	0.87	0.87	0.87	0.87	0.87	0.87
Box gearing fourth	0.9	0.9	0.96	0.96	0.96	0.96	0.96	0.96	0.96	0.96
Box gearing third	1.05	1.05	1.13	1.13	1.13	1.13	1.13	1.13	1.13	1.13
Box gearing second	1.29	1.29	1.42	1.42	1.42	1.42	1.42	1.42	1.42	1.42
Box gearing first	1.85	1.85	1.93	1.93	1.93	1.93	1.93	1.93	1.93	1.93
O/A top gear ratio	4.56	4.56	4.96	5.1	5.1	5.1	5.1	5.1	5.1	5.1
Fuel (l)	23	23	23	23	23	23.5	23	23	23	23
Box oil (cc)	–	–	1,600	1,600	1,600	1,700	1,700	1,700	–	–
Front wheel	3.0 x 18	3.0 x 18	3.0 x 18	2.75 x 18	2.75 x 18	3.0 x 18	3.0 x 18	3.0 x 18	3.0 x 18	3.0 x 18
Rear wheel	3.0 x 18	3.0 x 18	3.0 x 18	3.0 x 18	3.0 x 18	3.0 x 18	3.0 x 18	3.0 x 18	3.5 x 18	3.5 x 18
Front brake	dtls drum	dtls drum	dtls drum	dtls drum	dtls drum	disc	disc	disc	disc	disc
Rear brake	sls drum	sls drum	sls drum	tls drum	tls drum	disc	disc	disc	disc	disc
Front suspension	teles	teles	teles	teles	teles	teles	teles	teles	teles	teles
Rear suspension	dual shock	dual shock	dual shock	dual shock	dual shock	mono-shock	mono-shock	mono-shock	mono-shock	mono-shock
Wheelbase (mm)	1,316	1,316	1,331	1,331	1,331	1,316	1,316	1,316	1,321	1,321
Width (mm)	511	511	511	511	511	630	630	630	635	635
Length (mm)	1,930	1,930	1,946	1,946	1,946	1,935	1,935	1,935	1,935	1,935
Rake	27.5°	27.5°	27.5°	27.5°	27.5°	27.5°	27.5°	27.5°	–	–
Trail (mm)	90	90	90	90	90	75	75	75	–	–
Dry Weight (kg)	111	111	111	115	115	118	118	118	108	109

Appendix

750CC MOTORBIKE SPECIFICATIONS

MODEL	TZ750A	TZ750B	TZ750C	TZ750D	TZ750E	TZ750F
Year	1974	1975	1976	1977	1978	1979
Model code	409-000101	409-000361	409-100101	409-200101	409-200131	409-200197
Bore (mm)	64	66.4	66.4	66.4	66.4	66.4
Stroke (mm)	54	54	54	54	54	54
Capacity (cc)	694	747	747	747	747	747
Compression ratio	7.3	7.3	7.3	7.3	7.3	7.3
bhp	90	105	105	120	120	120
@ rpm	10,500	10,500	10,500	10,500	11,000	11,000
Ignition system	CDI	CDI	CDI	CDI	CDI	CDI
Ignition timing (mm)	2.00	2.00	2.00	2.00	2.00	2.00
Carburettor	Mikuni VM 34	Mikuni VM 34	Mikuni VM 34	Mikuni VM 34	Mikuni VM 34	Mikuni VM 34
Primary drive	2.6	2.6	2.6	2.6	2.6	2.6
Final drive	39/18	39/18	39/18	36/18	36/18	36/18
Box gearing sixth	0.75	0.75	0.75	0.75	0.75	0.75
Box gearing fifth	0.81	0.81	0.81	0.81	0.81	0.81
Box gearing fourth	0.89	0.89	0.89	0.89	0.89	0.89
Box gearing third	1.04	1.04	1.04	1.04	1.04	1.04
Box gearing second	1.2	1.2	1.2	1.2	1.2	1.2
Box gearing first	1.72	1.72	1.72	1.72	1.72	1.72
O/A top gear ratio	4.23	4.23	4.23	3.9	3.9	3.9
Fuel (l)	29	29	29	29	29	29
Box oil (cc)	1,500	1,500	1,500	1,500	1,500	1,500
Front wheel	3.25 x 18	3.25 x 18	3.25 x 18	3.25 x 18	3.25 x 18	3.25 x 18
Rear wheel	3.50 x 18	3.50 x 18	3.50 x 18	3.75 x 18	3.75 x 18	3.75 x 18
Front brake	disc	disc	disc	disc	disc	disc
Rear brake	disc	disc	disc	disc	disc	disc
Front suspension	teles	teles	teles	teles	teles	teles
Rear suspension	dual shock	dual shock	dual shock	monoshock	monoshock	monoshock
Wheelbase (mm)	1,407	1,407	1,407	1,390	1,390	1,390
Width (mm)	638	638	638	638	638	638
Length (mm)	2,037	2,037	2,037	2,014	2,014	2,014
Rake	27°	27°	27°	26°	26°	26°
Trail (mm)	–	–	–	97	97	97
Dry weight (kg)	157	157	157	152	152	152

Index

Adler, 12, 37, 42
Agostini, Giacomo,
  rider, 68, 69, 73, 78, 79, 81, 84, 85, 86, 87, 88, 90, 106, 107, 109, 110, 121,
  team manager, 155, 169, 170
AGV 200, Imola, 107
Aksland, Bud, 175
Aksland, Skip, 111
Andersson, Kent, 55, 56, 61, 62, 72, 73
Angel, Sonny, 39
Asama, 11, 12, 14, 16
Austrian GP, Salzbürgring, 81, 85, 92
Auto-bit, 9
Baker, Steve, 74, 88, 90, 109, 111, 118, 119
Bakker, 73, 74, 76, 99, 116
Baldwin, Mike, 169
Baldwin, Miles, 119, 120
Bartol, Harold, 128
Belgian GP, Spa Francorchamps, 19, 23, 26, 27, 131, 156
Bimota, 74, 76
Braun, Dieter, 67, 68, 86
Brauneck, Dave, 119
British GP, Silverstone, 94, 130
Brouwer, Ferry, 57, 62, 68
Cabton, 10
Cadalora, Luca, 135, 137, 139, 140, 180, 181
Cagiva, 179
Camillieri, Frank, 44
Carruthers, Kel,
  rider, 39, 56, 58, 59, 64, 97,
  tuner, 102, 106, 120, 134, 157, 159, 165, 169, 173
Castro, Don, 108
Catalina, 15
Cecotto, Johnny, 69, 72, 88, 90, 92, 93, 94, 110, 116, 118
Chilli, Pier-Francesco, 148
Cleek, Randy, 111
Crosby, Graeme, 153, 154, 155, 156, 158
Czechoslovakian GP, Brno, 27, 33, 68, 87, 90
Daytona, 44, 53, 59, 64, 106, 109, 115, 118, 119, 159
Degner, Ernst, 17, 21, 22
Disc-valve induction, theory of, 17

DKW, 10, 11
Dongen, Cees van, 20
Doohan, Mick, 177, 179
Duff, Mike, 26, 27, 28, 29, 31, 32, 52, 53
Du Hamel, Yvon, 53
Dulman, Boet van, 99, 100
Dunlop, 169, 174, 177
Dutch TT, Assen, 19, 23, 29, 59, 68, 85, 91, 94, 97, 99, 126
East German GP, Sachsenring, 27
Eguchi, Hideto, 102
Ekerold, Jon, 76
Elmore, Buddy, 44
Emde, Don, 60
Emmett, Sean, 183
Everett, Reg, 44, 45, 47
Fath, Helmut, 57
Femsa, 56
Findlay, Jack, 87, 114
Finnish GP, Imatra, 34, 73, 90
Fisher Gary, 60
Fontan, Marc, 100, 154
French GP, Clermont Ferrand, 19, 25, 27, 81, 88
French, Vince, 64, 66, 81, 116
Fuji, 9, 10, 11
Gardner, Wayne, 169
Garriga, Juan, 135, 137, 138
German GP, Nürburgring, 27, 92, 137
  Hockenheim, 34, 55, 73, 83
Gilera, 24
Godfrey, Tony, 23
Gould, Rod, 48, 53–55, 54, 56, 59, 61, 62, 81, 83
Hailwood, Mike, 31, 32, 34, 35, 78, 79
Harada, Tetsuya, 148
Harris Yamaha, 178, 182
Hasegawa, Toru, 23
Herweh, Manfred, 76
Hocking, Gary, 19
Hocking, Rick, 111
Hoeckle, 73
Honda, 10, 12, 16, 22, 34, 36, 78
  NSR500 (1984), 165–166
  NSR500 (1992), 178
  RC142, 16, 22
  RC160, 16, 37

189

Index

RC162, 39
RC163, 79
RC165, 26
RC166, 31, 32
RC170, 78, 79
RC174, 31
RC181, 79
RS500, 177
Hummel, 127, 128, 129, 130, 137
Isle of Man, 19, 23, 28
Italian GP, Monza, 26, 31, 57
    Imola, 55
Ito, Fumio, 15, 19
Ivy, Bill, 29, 32, 33, 34, 36, 41, 42
Japanese GP, Suzuka, 22
Kaarden, Walter, 17
Kanaya, Hideo, 66, 69, 80, 81, 83, 115
Kanemoto, Erv, 108, 116, 118, 119, 163
Katayama, Takazumi, 72, 73, 74, 92, 93, 148
Kocinski, John, 135, 137, 139, 140, 141, 175
Kröber, 56, 57, 73
Kurth, Rudi, 72, 73
Lansivouri, Tepi, 65, 66, 68, 84, 85
Lavado, Carlos, 123, 125, 126, 129, 130, 132, 133, 135, 137
Lawson, Eddie, 159, 161, 163, 165, 167, 169, 170, 171, 172, 173, 174
Lectron, 123
Loudon, 58, 60
MacKay, Mac, 84, 86
Mackenzie, Niall, 173, 175, 183
Maekawa, Mike, 93, 151
Magee, Kevin, 170, 172
Mamola, Randy, 169, 170
Mann, Dick, 44
Manxton, 74
Masuko, 14
Matsui, Takasi, 80
Meguro, 13
Michelin, 166, 171, 174, 175, 177
Middelburg, Jack, 99, 149, 150
Miyashiro, 14
Monarch, 10
Monoshock, theory behind, 69–70
Morini, 23
Mortimer, Chas, 61, 62
Motohashi, 80
MZ, 17, 21
Nixon, Gary, 44, 47, 74, 108, 116, 118
Noguchi, Taneharu, 16, 19, 37

Nurmi, Pekka, 74
Öhlin, 159, 172, 177, 180
Oishi, 11, 14, 19
Ontario, 58, 106
Palomo, Victor, 116
Parrish, Steve, 100
Pierce, Ron, 53, 74
Pointer, 10
Pons, Patrick, 119
Powervalve, theory behind, 91
Provini, Tarquinio, 24, 25
Quaiffe, Ron, 57
Rainey, Wayne, 117, 127, 170, 172, 174, 175, 176, 177, 180, 181, 182
Read, Phil, 24, 25, 26, 27, 28, 29, 31, 32, 33, 34, 36, 41, 55, 57, 58, 81, 83, 85, 87
Redman, Jim, 22, 23, 24, 25, 32
Reed valves, theory behind, 164
Reynolds, John, 183
Robb, Tommy, 22
Roberts, Kenny, 60, 68, 74, 75, 90, 92, 93, 94, 95, 96, 106, 107, 109, 110, 112, 113, 115, 119, 121, 135, 149, 150, 151, 152, 153, 154, 156, 157, 158, 159, 160, 161, 168, 169, 170, 173, 177, 180
ROC Yamaha, 178, 179, 181, 182
Romero, Gene, 106, 108, 110
Ruggia, Jean-Paul, 136, 137, 139, 178
Saarinen, Jarno, 58, 60, 62, 63, 64, 66, 78, 80, 81, 82, 83
San Marino GP, Imola, 161, 162
Sarron, Christain, 100, 123, 125, 126, 128, 129, 168, 170, 171, 178
Schlögel, Sepp, 68
Schmidt, Jochen, 145, 146, 148
Schwantz, Kevin, 177, 181
Schwerma, Doug, 111
Sheene, Barry, 85, 88, 90, 92, 94, 100, 114, 153, 154, 155, 156
Showa, 10, 11, 14, 17
Sinclair, Mike, 169
Smrz, Gregg, 119
Spanish GP, Jarama, 58, 72, 92
Spencer, Freddie, 129, 130, 155, 161, 163, 166, 167, 168, 169, 173, 175
Spondon, 73, 74
Sunako, Yoshikazu, 19, 23
Suzuki, 10, 101, 179
    RZ65, 26, 29
    RG500, 97, 149–150, 177

Index

Swedish GP, Anderstop, 88, 92
Taira, Tadahiko, 132, 134
Tass, 14
Taveri, Luigi, 33
Tilbury, Trevor, 72, 73, 97, 155
Tohatsu, 10
Ulster GP, Belfast, 27, 33
USA GP, Daytona, 27
Vesco, Don, 119
Vincenzi, Giuseppe, 56
Warburton, Brian, 42
Watase, 11
Wimmer, Martin, 124, 125, 126, 127, 129, 130, 131, 132, 134, 135, 136, 137
Yamaha,
   350-3, 72, 74
   GL750, 101
   OW16, 66
   OW17, 66, 68
   OW19, 64, 68, 79, 80, 80–81
   OW20, 83, 84
   OW23, 85
   OW26, 87, 88
   OW29, 88
   OW31, 114, 115, 117
   OW35, 88–89, 90
   OW45, 93–94
   OW48, 95–96
   OW48R, 97
   OW54, 150, 151, 154
   OW61, 156–157
   OW70, 159, 163
   OW76, 164–165
   OW81, 166, 167, 168
   RA31, 33–34, 35
   RA41, 17–18, 14
   RA97, 25, 28, 29
   RD05, 28, 29–31, 35
   RD48, 17, 18–21
   RD56, 22–25, 24, 26, 44
   TD1, 39, 40
   TD1A, 41–42
   TD1B, 42–45, 43
   TD1C, 45–48, 46
   TD2, 49–51
   TD2B, 57
   TD3, 59, 102
   TR1, 51–52
   TR2, 52–53, 55
   TR2B, 57
   TR3, 59, 60, 62
   TZ250, 63, 66
   TZ250A, 1973, 65
   TZ250A, 1991, 141, 142
   TZ250B, V-twin, 142–144, 145
   TZ250C, 69, 70–72
   TZ250D/TZ350D, 72
   TZ250D, V-twin, 144
   TZ250E/TZ350E, 73
   TZ250F, 74–75
   TZ250G, 76–77, 101
   TZ250H, 122–123
   TZ250J, 125
   TZ250K, 125
   TZ250L, 126, 127
   TZ250M(1992–3), 145, 146, 147
   TZ250N, 129
   TZ250S, 130
   TZ250T, 135
   TZ250U, 135–136, 136
   TZ250W, 137–139
   TZ350, 62–64
   TZ350C, 69, 70
   TZ350F, 75–76
   TZ350G/H, 76
   TZ500G, 97–99, 121
   TZ500H/J, 100
   TZ700/750A, 64, 103–106, 105
   TZ750B, 109
   TZ750 Miler, 111–114
   TZ750D, 117
   TZ750E/F, 118
   YA1, 10–12, 11
   YDA/B, 13–14
   YDS1R, 16, 37–39, 38, 41, 42
   YES1R, 16
   YE1, 16
   YZR250 (1985), 130, 131
   YZR250 (1986), 132–134
   YZR250 (1987–9), 134, 136, 138
   YZR250 (1990), 139
   YZR500 (1986), 167
   YZR500 (1988), 171
   YZR500 (1989), 173, 174
   YZR500 (1990), 175
   YZR500 (1991), 176, 177
   YZR500 (1992), 179, 180
Yugoslavian GP, Opatija, 72